Stay Healthy, Stay Youthful: The Science of Living to 150

Cutting-Edge Longevity Science for Reversing Aging and Living a Healthier, Longer Life

HEALTH AND LONGEVITY MASTERY SERIES
VOLUME ONE

By Tad Sisler

To my father, Maynard. You were an outstanding physician and healer. You cared; you listened; you left the world in a better place.

Your continuing thirst for knowledge and excellence with compassion created a fire in my childhood that burns to this day. Thank you for pushing me to think outside the box and become a better person. Godspeed, my loving father.

TABLE OF CONTENTS: PAGE

FOREWORD

My father, **Maynard Lee Sisler, M.D., F.A.C.P.**, instilled a love of science and medicine in me from the moment I could walk and talk. In addition to being a board-certified physician, he was a Renaissance man. Dad had read all the classic books early and imparted as much wisdom as he could muster to me as I grew. At age ten, I memorized **Shakespeare's** soliloquy from *Hamlet* and read many of the classics myself, including some within his vast library of books on physics and medicine. Dad was an internist, a Captain in the United States Navy, and became Chief of Medicine at the largest Army-Navy hospital in the world in Corpus Christi, Texas. Later, he became a *Fellow* in the **American College of Physicians** and a member of London's **Royal Society of Medicine.**

Maynard Lee Sisler, M.D., F.A.C.P
Source – Sisler Private collection

The world of medicine in my father's era was one of medical miracles: the discovery of antibiotics and the polio vaccine among them. At the same time, modern medicine began to move away from the centuries-old holistic approach and became more dependent upon pharmaceuticals and treating rather than preventing disease. Prescribing medicine rather than looking at and preventing a root cause became a norm. Many doctors viewed patients as customers.

I remember my dad being hounded by drug salespeople who lavished him with pens, notepads, and other gifts he subsequently brought home as 'prizes' for me. I knew the names and interactions of many drugs before I was eight. Still, my father was a careful, thoughtful physician, doing his best to treat and heal his patients with the tools of his era.

At that point in time, there was a much larger market for pharmaceutical drugs than for healthy lifestyles, yet still my dad cautioned against over-prescribing, long before it was fashionable.

My father was one of the first to implement the medical miracles of the early-to mid-20th Century. It's time to go back to medical miracles. I strongly believe we are headed in that direction now.

Between my father's years running U.S. Naval Hospitals in Corpus Christi and San Diego, he would frequently take me on rounds at the hospitals. Many of his patients were old World War II retired Admirals with a series of maladies brought on by challenging living conditions and societal norms during their lifetimes, like smoking and heavy alcohol use. Dad would show me advanced cases of diabetes, or gout, or how heart disease or cancer can debilitate a person. Rather than being 'grossed out,' I was fascinated.

My father had subscriptions to all the leading medical journals. As a child, I would read articles in many, including the *New England Journal of Medicine*. In an era when many people had *Life* or *Vogue* magazines in their homes, I had medical journals to thumb through. Of course, much of the information was above my head, but my father showed me how to navigate medical information to discern which studies or trials may have been paid for, or possibly even manipulated by the pharmaceutical companies to promote a new drug. Even a child can see propaganda when it is pointed out to him or her.

My father scorned the medical profession for seeing the body as many separate parts rather than a unified system. Aside from his time in the Navy, he hated that 'good' insurance was only available for those who could afford it. In fact, when he retired from the Navy and went into private practice as a small-town country doctor in Missouri (his lifetime dream), he would accept chickens or watermelons or other barter for his fees from struggling farmers on occasion.

Later in his career, my father railed against the fact that big pharma uses society as guinea pigs and inflates the cost of life-saving drugs, and that the medical establishment treats symptoms before issues. He told me that all a physician must do is to listen to their patient, and they will find out so very much before they ever need to prescribe anything. My father became a favorite teacher and mentor for interns, receiving many awards and accolades. His devotion to medicine fascinated me, and as a young child I believed he had healing powers in his noble profession. My love for medicine only grew through time, but my career was destined to go in another direction.

My mother, **Elaine Witt Sisler**, was a concert pianist, teaching me from a very young age how to play piano. She had been a child prodigy, giving her first recital at 5, and performing as a soloist for the *Chicago Symphony* at 17. Her passion for performing became my own. At some point, my love of music trumped my love of medicine, and I devoted my life to music and the arts.

The die was cast, however, and my father's insatiable quest to, as he put it, *"always make sure your reach exceeds your grasp"* caused me to take the Socratic approach through life, asking more and more questions, and continuing my studies in medicine and science.

Tad Sisler's Mother Elaine Witt Sisler
Source – Sisler Private Collection

At the age of 22, I was working a grueling, six-day-a-week schedule driving two hours in each direction up and down a Nevada mountain road from Reno to South Lake Tahoe to work, setting up my equipment, working a six-hour gig performing at **Harvey's,** breaking down, and driving back late night to my waiting family. To stay awake while driving home, I would open the car windows and stick my head outside the window for moments in the frigid cold, and I listened to talk radio. The great broadcaster **Larry King**, who later became my friend, had a national late-night radio show interviewing all the iconic people of the time. One night, **King** interviewed the great doctor **Michael DeBakey**, who had performed the first heart transplant. **Dr. DeBakey** talked about heart health through cardiovascular exercise and a good diet.

He also mentioned that, if you're a runner, you should consider power walking instead of running after you reach the age of forty because it is better on your heart, and your knees and hips won't break down as quickly as they would, should you continue running.

At the time, I wouldn't have dreamed that, at some point in my lifetime, new stem-cell therapies might allow us all to regenerate our knees, hips, or shoulders without worry of degeneration or painful surgeries.

On another night, **Larry King** was interviewing a scientist (I wish I could remember his name!). The scientist said that, throughout history, species that did not evolve to expand their brain capacity over a period of 75,000 years were doomed to go extinct.

6

He mentioned that the human brain has not expanded its capacity over this period. Still, he believed that because we invented computers, we could externally accomplish this, and computers (as a higher brain) would save the species. This concept mirrors the words of **David Gerrold:**

"In the entire history of the human species, every tool we've invented has been to expand muscle power. All except one. The integrated circuit, the computer. That lets us use our brain power."

Sometimes, today, when I look at what could come of military robotics, drones, and other terrible inventions, I fear a *Terminator* scenario! But I do believe the scientist had an excellent point, and these inventions will naturally expand our lifespan through new and accelerated discoveries. Nano computers in the bloodstream cleaning out plaque, reducing disease or allowing for more brain function may already be on the horizon. Progress is always scary at first.

Larry King and Tad Sisler
Source – Sisler Private Collection

Years later, I was driving home late at night from another gig, listening to another radio show, and the host mentioned that he believed the first person who would live to be 150 had already been born. I was shocked to hear this because the person who had lived the longest in my own family was my grandmother **Audrey Athey Sisler**, who made it to 94.

More recently in 2025, **Bryan Johnson**, an entrepreneur and venture capitalist, claimed that he was working on the concept of living forever while studying anti-aging and infusing his blood with his teenage son's plasma. **Dr. Aubrey de Grey** is an English biomedical gerontologist known for his view that medical technology may enable human beings alive today not to die from age-related causes, suggesting that there may be someone alive today who will live 1,000 years. Although this seems implausible by today's standards, many scientists believe that, with the advent of quantum computing, if we can make it another five years, we will have the technology to easily live to the ripe old ages of between 120 and 140.

Although today, more centenarians are alive on earth than at any other time in recorded history (573,000 estimated in 2024 according to *Worldometers*, which is only 0.007% of the total global population), according to the *Gerontology Research Group*, there may be only between 150 and 600 supercentenarians — individuals aged 110 years or older — alive today.

So, with such a small current sample size, how could anyone claim that people who are already born will make it to 150?

As I researched this topic, reading excellent new books I would highly recommend, including *"Lifespan"* by **David A. Sinclair** and *"Life Force"* by **Tony Robbins**, I was amazed at the amount of longevity research and development happening right now by leading scientists worldwide. Having myself won a coveted **Reader's Favorite Award** as the author of a biography I wrote on a famous trumpeter; I'm always searching for books that captivate me and help me stretch my imagination as far as possible. Although each of these books (and others I found) are precious sources of information regarding aging, I haven't seen a book yet that incorporates cutting-edge science with a rounded view, including natural remedies throughout the ages, psychological and mental health effects on aging, important nutraceuticals and supplements and possible contraindications, lifestyle strategies based upon history and genetics as well as science, environmental toxins to avoid, or a blueprint on what I could do today and every day to have the greatest chance of making it to 150. I do my best to achieve this within these pages. I've admittedly been frustrated that so many advancements are on the horizon. Yet, there is not much we can do at this moment besides eating right, exercising, taking a few new supplements, and getting enough sleep. But it's a good start. Through this book, I project time limits for when we can expect new developments to occur (to the best of my ability). At the end of the book, I'll point you towards continuing research.

—

A study by *Stanford Medicine* reveals the invention of "smart toilets", disease-detecting toilets that can sense multiple signs of illness through automated urine and stool analysis. I read somewhere else that scientists are working on the idea of a futuristic 'smart house' where your toilet analyses your urine and stool, and by the time you get to the kitchen, your refrigerator is ready to dispense a perfect concoction of supplements, nutraceuticals and pharmaceuticals to keep you in optimum health for the day. The future has endless possibilities!

As a filmmaker, I won a coveted **Telly Award** for my documentary, *"Journey to an Extraordinary Life."* Delving into what makes successful people tick was an eye-opener because I found a thread of commonality in every legendary person I interviewed.

We are all connected in this earthly existence in ways we never dreamed of. Within all of us are the seeds for success, health, and longevity. Artificial Intelligence is bringing a new world of discoveries and changes, but our humanity must always be in the forefront as we incorporate nanobots literally into our bodies to extend our health and longevity. I believe the master inventor and author **Ray Kurzweil** when he says:

"By the time we get to the 2040s, we'll be able to multiply human intelligence a billionfold. That will be a profound change that's singular in nature. Computers are going to keep getting smaller and smaller. Ultimately, they will go inside our bodies and brains and make us healthier, make us smarter."

Ray Kurzweil
Credit – Wikimedia Commons

Yet, how do we navigate through *"the slings and arrows of outrageous fortune,"* calamities that may happen at any moment, disease, accidents, and acts of nature? How do we conquer drug addiction, violence, obesity?

My loving sister **Judith Sisler Pedro** was a Registered Nurse with a master's degree. **Judy** was an Officer in the *United States Air Force*. She was brilliant and could teach any medical concept as if reciting from a textbook. Yet, with all the knowledge and insight she had, at 64 she suffered a tick bite at a festival in Wyoming and died three weeks later from Rocky Mountain Spotted Fever. By the time the illness was diagnosed, it was too late. I was devastated at the sudden, unexpected loss of my sister, an otherwise healthy person. Some things are beyond comprehension or explanation. Life is a crap shoot. If you believe in life after death, some things must be left to God, the Universe, or possibly fate, yet the most we can do is to act now to put the odds in our favor.

Tad Sisler's Sister, Judy Pedro
Source – Sisler Private Collection

Before my sister **Judy** died, we would take long walks and talk about the meaning of life. In one conversation, she predicted that, when we die, so much will be revealed to us that we just cannot see or comprehend now because we're presently locked up in the limited physics of Earth. I wish I could talk to her today about the miracle we are about to witness with quantum computing. Within the next ten years, we are going to see about four hundred years of progress, and much of this will be in the area of life extension.

My father's brother, my Uncle **Bill Sisler**, said, *"Later in life, I realized that for every year I had worked and run hard trying to squeeze two years into every year of my youth, the result each year was to subtract a year from the end of my life."* It's okay to work hard and play hard, but at the same time, we must do our best with what we've got to stay healthy and vital yet learn to pace ourselves and cut down on stress whenever we can. In 2008, **Ray Kurzweil** also prophesied that new advances in longevity will be coming exponentially during the 2030s if we can stay alive until then. Buckle up!

THE HAPPINESS QUOTIENT
I will reiterate many times in this book the idea that attitudes and emotions will dictate your happiness quotient in life and contribute heavily to longevity. What is the meaning of a long life without some level of happiness? And how does one become happy or sustain contentment and joy? For more on this subject, check out my book **The Science of Positive Thinking: How Mindset, Daily Habits, and Emotional Well-being Can Add Years to Your Life.**

Elon Musk recently posted on the *X* platform that *"We are on the event horizon of the singularity."* The singularity is defined as a hypothetical time when machines surpass human intelligence, leading to a new era of rapid technological evolution. Hypothetical or not, the era of quantum computing has arrived, and we are going to experience several hundred years of innovation within the next decade, upending everything we know about the human experience. So, what will be left when machines take over all basic chores and most jobs? Machines may be vastly more educated than us, but the human experience will remain unique, and basic themes of family, community, love, and kindness will become even more important as we learn to live our lives differently. Add longer lifespans and healthspans to that, and you have the recipe for an amazing century ahead. Life, liberty, and the pursuit of happiness should be everyone's goal.

The **World Happiness Report** (and other global well-being studies) consistently places countries like **Finland, Denmark, Norway, Iceland**, and **Switzerland** at or near the top of happiness rankings. These nations share several core characteristics—strong community bonds, social support, work-life balance, and robust public services—that contribute to overall contentment. Below are **ten of the most important "keys to happiness"** gleaned from these happiest places, emphasizing community, family, harmony, and quality of life.

STRONG SENSE OF COMMUNITY AND SOCIAL SUPPORT

People in happier countries report feeling supported by friends, neighbors, and local organizations. This profound social trust extends into daily life—whether borrowing a bit of sugar from your neighbor or relying on a community group in times of crisis.

In the Nordic nations, communal activities and clubs (e.g., sports and cultural associations) are common, which fosters strong social networks and reduces isolation.

EMPHASIS ON RAISING CHILDREN AND FAMILY WELL-BEING

Societies that invest heavily in childcare, education, and parental leave allow parents to balance work and family. This reduces stress and promotes healthier childhood development. Generous parental leave policies in Scandinavia and subsidized childcare ensure parents can bond with their children without fear of losing income or job security. Nurturing a healthy, happy family is the key to reducing violence and crime in society.

WORK-LIFE BALANCE

All work and no play makes Jack the dull boy. Jobs allowing flexible hours, enough leisure time, and vacation days help people maintain their mental and physical health. Chronic overwork often leads to burnout, which diminishes overall happiness. Remember the adage, though, that if you love what you do, you don't work a day in your life. Choose a profession that makes you happy if you can.

In Finland and Denmark, the standard full-time workweek often includes strong protections for time off, and it's culturally accepted to leave the office on time to be with family or pursue personal interests.

TRUST AND GOOD GOVERNANCE

High trust in government and fellow citizens correlates strongly with happiness. When institutions are transparent and fair, people feel secure and less stressed. Low corruption, access to working government services, and social welfare programs in many Northern European countries lead to a sense of security— people believe their taxes are used responsibly for the common good. The United States may be moving back in the direction of accountability and transparency in government, which should bring about a greater sense of general happiness in America.

EQUALITY AND SOCIAL SAFETY NETS

Broad social safety nets (e.g., access to healthcare, unemployment benefits, pension systems) cushion life's hardships. We must be careful, though, not to spoil this with overtaxation and overregulation. A strong baseline of equality helps reduce stress and resentment, fostering social harmony. Love your neighbor and learn to understand them rather than to fear or loathe your cultural differences. Let go of anger and hatred, and don't let other people's actions get under your skin. Remember the transformative words of **Miguel Ruiz** in his book *The Four Agreements:*

"Don't take anything personally."

The gap between the richest and poorest is smaller in many of the happiest countries, reducing social tensions and improving overall community well-being. Mostly, though, I believe that the idea of rich vs. poor fosters resentment and could work against our quest for more happiness. Learning to be happy with what we have is a challenge and a blessing.

HIGH-QUALITY EDUCATION

Accessible education systems nurture children's intellectual, emotional, and social growth. Education also provides opportunities for upward mobility, contributing to a sense of hope and fairness. The United States learned a valuable lesson that vast funding is not always the answer. Training brilliant teachers who care is a much better concept than throwing money at a problem. Throw money at the finest teachers' salaries instead based upon merit.

Finland's education system is world-renowned for its focus on student autonomy, creative thinking, and minimal standardized testing—leading to high satisfaction among students and teachers alike.

HEALTHY LIFESTYLE AND CONNECTION TO NATURE

Regular physical activity, emphasis on outdoor recreation, and access to green spaces contribute to better mental health. Spending time in nature lowers stress and promotes happiness.

Nordic countries encourage biking or walking for daily commuting, have abundant parks and recreation areas, and even prioritize **"friluftsliv"** (the Norwegian concept of outdoor life).

CULTURE OF WORKABLE BALANCE BETWEEN AUTONOMY AND COMMUNITY

People thrive when they feel individual freedom (to make choices and pursue passions) and strong belonging (that others have their back). Balancing personal autonomy with collective well-being reduces stress.

In Denmark, the concept of **"hygge"** (coziness and comfortable conviviality) combines personal comfort with social bonding, reflecting personal autonomy and community spirit.

PURPOSE, MEANING, AND LIFELONG LEARNING

What about the people who simply don't want to live this long? I could not believe someone would not want to live an extended life if they were healthy and generally happy until I ran into a friend in her seventies and told her about this book. She looked at me almost confusedly and said, *"Oh, no, I'll be happy to live to 90 and then I want to leave the planet. I've had no children, and things aren't the same now as when I was younger."* She liked life better then, she was unhappy with the current political situation as she viewed it... and even when I asked her if she thought she might think differently when she becomes 90 and still feels great... I could not sway her strong conviction that she just absolutely did not want to even consider an extended lifespan. I must respect this way of thinking, and I know that everything changes around us as we evolve and progress. Sometimes, life doesn't turn out the way we expected, but we can still find meaning in our interests, family, volunteering, helping others, or a hundred other things. If I'm given the chance to live a very long life and to be healthy, I would feel almost obligated to make the most of this gift. Of course, I pray for the safety of my family daily, and I know that we all experience tragedies and low points, but as my blues musician friend said, *"If we didn't feel the extreme lows, we wouldn't appreciate the highs so much."*

Shakespeare's soliloquy from *Hamlet*, which I memorized at 10, is a deeply personal revelation by a haunted man contemplating suicide. He admits at some point that **"conscience doth make cowards of us all,"** deciding that his fear of what lies beyond death is stronger than his desire to end his life. At some point in our lives, we will all experience crippling sorrow. How will we deal with this when it happens? How did my grandmothers find the strength to deal with their crippling sorrows? I believe the answer is courage. **Sir Winston Churchill** said:

"Without courage, all other virtues lose their meaning."

When my wife died, my father called me and reminded me of the great quote by **Albert Camus**, **"In the midst of winter, I found there was, within me, an invincible summer."** Dig down deep into the strength of your soul, your shining light, and find your own invincible summer. Even the cowardly lion in *The Wizard of Oz* found that he always had courage; he had just lost sight of it for a time. And that's ok. There are concrete steps you can take to pull yourself out of despair. Don't stay there. Reinvent yourself. Always find something to look forward to.

People who find meaning in their work, hobbies, or volunteer efforts report higher life satisfaction. Communities that encourage continuous skill-building and personal growth see lower rates of depression.

Adult education courses and funded opportunities to learn new skills are widespread in many European countries, helping people stay intellectually engaged beyond traditional schooling years.

CULTURAL NORMS OF GRATITUDE AND OPTIMISM
A cultural mindset that prioritizes thankfulness and celebrates simple pleasures (like a shared meal or a good conversation) reinforces positive thinking.

In Iceland, despite challenging weather and long winters, people often talk about the beauty in everyday moments and support each other's pursuits—resulting in a resilient, positive community outlook.

To summarize, happiness is **multifaceted**: it depends on social structures (like effective governance and social safety nets), personal lifestyle factors (like work-life balance and outdoor activities), and cultural values (like trust, gratitude, and community harmony). The **happiest countries** reflect these **ten core principles** in policy and daily life. Adopting some of these keys can lead to a more content, harmonious, and supportive environment if you're seeking to enhance well-being—on a personal level or within a larger community.

Working towards a goal of general contentment will add quality years to your life. And mostly…although the factors I've just mentioned are known to play a part in general happiness, truly it's just a simple choice. You can wake up each day deciding to be happy or not. It's all up to you. And I promise, the pursuit of happiness will contribute greatly to the chance you may live to be 150. A 2023 study in *The Lancet* found that individuals with high optimism and happiness had a 15% lower risk of all-cause mortality over 10 years, linked to better heart health and lower stress.

You'll see other recurring themes throughout this book… Reduce or eliminate inflammation; sleep enough; eat right; exercise; take supplements; consider intermittent fasting. All of these are just common-sense suggestions for optimal health, and you could imagine that incorporating these into your life could absolutely increase your potential for longevity. You might get tired of repeatedly seeing these recurring themes throughout the book, but take heed: these are the building blocks for everything else.

I hope and believe that someone will open this book a century from now and laugh at what we still didn't know as we moved into the second quarter of the 21st century. But maybe that same person will marvel that we had all the building blocks in place for advancements in longevity and age reversal that they will hopefully take for granted by then. I sincerely hope you live to be 150; when you do, the world will be different. Meanwhile, do your best to spread love, kindness, and joy (you will find that positive attitudes and emotions are part of getting there anyway), and I hope to see you then! First, let's find out how we've already come so far.

"Youth has no age." – Pablo Picasso

EXTREMELY IMPORTANT – READ THIS BEFORE PROCEEDING

MEDICAL DISCLAIMER

The information provided in this book is for informational purposes only and is not intended to replace the advice, diagnosis, or treatment provided by a licensed medical professional. The author and publisher are not medical practitioners and do not claim to offer medical advice. Readers are advised to consult with a qualified healthcare provider before starting any new health regimen, including but not limited to dietary changes, exercise programs, use of supplements, or other health-related practices discussed in this book. Some of the ideas or techniques described in this book may not be viable or available, and others are in development and not yet tested. The methods, techniques, and practices described in this book may not be suitable for all individuals, and their effectiveness and safety can vary based on personal health conditions and circumstances.

No guarantees or warranties of any kind are made regarding the accuracy, completeness, or suitability of the information provided. The reader assumes full responsibility for any actions taken or decisions made based on the contents of this book. The author and publisher disclaim any liability for any injury, loss, or damage caused directly or indirectly by the use or misuse of the information contained herein.

LEGAL DISCLAIMER

This book is presented for informational and educational purposes only. It is not intended to provide medical, legal, financial, or any other form of professional advice. The content is based on research, opinions, and sources believed to be reliable at the time of publication; however, the author and publisher make no representations or warranties regarding its accuracy, completeness, or applicability to any particular individual or circumstance. The reader is solely responsible for any decisions made or actions taken based on the information in this book. By reading this book, the reader agrees to hold the author, publisher, and any affiliated parties harmless from any and all claims, demands, liabilities, or damages arising directly or indirectly from the use, misuse, or application of any content within. This book does not establish a professional relationship between the author, publisher, and reader. Readers should always seek the advice of a qualified professional for specific concerns, issues, or decisions. Any reference to products, practices, or methods does not constitute an endorsement or guarantee by the author or publisher. The content, including any discussions of supplements, therapies, or emerging scientific developments, is subject to change as new research and evidence become available. The author and publisher disclaim any responsibility for errors or omissions and are not responsible for any consequences arising from the use of this book.

Credit: rawpixel.com

INTRODUCTION
THE TIMELESS PURSUIT OF YOUTH
HISTORICAL PERSPECTIVES ON AGING AND IMMORTALITY

S ince the dawn of civilization, we have been captivated by the allure of eternal life. Myths and legends from every corner of the world echo this desire, telling great stories of mystical fountains, elixirs, and divine interventions granting immortality or prolonged youth.

The Bible states that Methuselah lived to 969 years, and Jared lived to 962. If it's difficult for you to accept the literal interpretation of these unusually long lifespans, they could be symbolic representations of wisdom, perhaps, or ancient numerology may have influenced how ages were recorded. At this writing, the longest-lived documented human in our time is Jeanne Calment, from France, who lived to be 122, and died in 1997.

Imagine the possibilities of human discovery and wisdom, if we are all given the opportunity to live longer! What more would an Edison have invented, or an Einstein postulated? How much better of a pianist might Art Tatum have become if that's even possible? Let's dig deep into the rich tapestry of historical perspectives on aging and immortality, and on the way, we'll explore ancient myths, early medical practices, and the philosophies of influential thinkers who shaped our understanding of health and longevity. Then, we'll start thinking outside the box.

"We cannot solve our problems with the same thinking we used when we created them." – Albert Einstein

Albert Einstein
Credit – Wikimedia Commons

ANCIENT MYTHS AND THE PURSUIT OF ETERNAL LIFE
The Fountain of Youth and the Elixir of Life are among the most enduring legends symbolizing humanity's desire to escape the confines of mortality.

The **Fountain of Youth**, a spring that makes someone who drinks or bathes in its waters young again, appears in various cultures—from the stories of **Herodotus** about the Ethiopians' magical fountain to the tales of Spanish explorer **Juan Ponce de León** searching for it in the New World.

Interestingly enough, a group of scientists in the late 20th century, including **Stanley Skoryna** and **Georges Nogrady**, took soil samples from Easter Island (a lake on Easter Island was rumored to have rejuvenating properties similar to a **Fountain of Youth**), and discovered rapamycin, an amazing compound we'll delve more deeply into in later chapters. Here's another perspective:

"There is a fountain of youth; it is your mind, your talents, the creativity you bring to your life and the lives of people you love. When you learn to tap this source, you will truly have defeated age." – Sophia Loren

Sophia Loren
Credit- Wikimedia Commons

Similarly, the **Elixir of Life**, often associated with alchemy, was thought to be a potion that would grant the drinker eternal life or youth. Alchemists like **Paracelsus**, a Swiss physician and philosopher of the Renaissance, dedicated their lives to discovering life-extending substances. **Paracelsus** revolutionized medicine by introducing chemistry and challenging traditional medical practices. He believed in the harmony of the body and spirit, and that understanding nature's secrets could unlock the mysteries of life and death.

PHILOSOPHIES OF AGING IN ANCIENT CIVILIZATIONS

EGYPTIAN PERSPECTIVES

Ancient Egyptians looked at aging through the lens of spirituality and the afterlife. They practiced meticulous mummification to preserve the body for the soul's journey beyond death.

Egyptian medical knowledge, recorded in texts like the **Ebers Papyrus**, included remedies and practices to maintain health and prolong life. They believed a balance of physical and spiritual well-being was needed to achieve longevity.

GREEK INSIGHTS

The Greeks approached aging with a blend of philosophy and empirical observation. **Hippocrates**, the "Father of Medicine," emphasized the importance of diet, exercise, and environmental health factors. He advocated for the body's natural healing processes and introduced the concept that lifestyle choices directly impact our lifespan. ***"First, do no harm"*** is a quote attributed to **Hippocrates,** although not in the *Hippocratic Oath* as many believe.

Pythagoras, more renowned for his contributions to mathematics, also founded a school that viewed the body and soul as interconnected entities. He promoted dietary restrictions and mental discipline as pathways to a longer life.

CHINESE WISDOM

Ancient Chinese civilization offered profound insights into aging and longevity through philosophies like **Taoism** and **Traditional Chinese Medicine (TCM)** practices. Pursuing harmony with the **Tao**, or the natural order, was believed to lead to health and longevity.

Ancient Chinese developed techniques like acupuncture, herbal medicine, and qigong to balance the body's energies. The legendary **Shen Nong**, the "Divine Farmer," was said to have tasted hundreds of herbs to discover medicinal properties, laying the foundation for herbal medicine to prolong life. I wonder how many times he got sick eating the wrong plants!

EARLY MEDICAL PRACTICES TO PROLONG LIFE

AYURVEDIC TRADITIONS

In India, the sages **Charaka** and **Sushruta** composed foundational texts of Ayurveda, one of the world's oldest holistic healing systems. **Charaka's** writings focus on internal medicine and the principles of maintaining health

19

through balance, while **Sushruta** is known for his surgical techniques and detailed anatomical knowledge. Ayurveda emphasizes a balance of the body, mind, and spirit, advocating for personalized treatments and preventive care to enhance longevity.

Their texts also mention water purification methods, like boiling, exposure to sunlight, and filtration through sand and charcoal, underscoring the ancient understanding of clean water's role in health.

WATER TREATMENT SYSTEMS IMPACT ON LONGEVITY

Another great **Larry King** interviewee claimed that water treatment plants were the single most crucial reason human life expectancy jumped dramatically from the beginning of the 20th century until today. Access to clean water has been a fundamental factor in human health and longevity throughout history. The development of water treatment systems marks a significant milestone in public health.

ANCIENT WATER PURIFICATION METHODS

From the very beginning, people understood how important clean water was. Ancient writings from India (around 2000 B.C.) describe boiling water, filtering it through sand or gravel, and leaving it in sunlight to purify it. In Egypt, around 1500 B.C., they used a substance called alum to make dirt and other particles settle to the bottom.

Hippocrates, the famous Greek physician, even invented something called the **"Hippocratic sleeve."** It was basically a cloth bag used to filter rainwater or boiled water, removing sediments so it tasted better and was safer to drink.

INNOVATIONS IN THE 17th THROUGH 20TH CENTURIES

The path to modern water treatment took huge steps in the 19th century. One key figure was **Dr. John Snow,** a British physician who traced a cholera outbreak in 1854 to one contaminated water pump on Broad Street in London. By mapping where people got sick, he proved that water could spread disease, paving the way for modern public health methods.

Before that, in the early 17th century, **Sir Francis Bacon** tried to remove salt from seawater by filtering it through sand. Though it wasn't an instant success, it inspired future experiments. Then, in 1829, **James Simpson** introduced slow sand filtration in London, which did a great job removing contaminants.

By the early 1900s, chlorination became a standard practice to disinfect water. **Dr. John L. Leal** first used chlorine continuously in Jersey City, New Jersey, in 1908. This simple treatment cut down waterborne diseases—like typhoid fever—almost overnight.

IMPACT ON LONGEVITY

Many people wonder: "How did our life expectancy shoot up so much in just one century?" The truth is, it happened over many centuries of trying things out, suffering through diseases, and fine-tuning what works. But the development of water treatment stands out. Once we could clean up our water supply, diseases like cholera and dysentery dropped sharply, especially in developed nations.

Safe water also boosted public health overall—fewer babies died in infancy, and more people lived longer. Cities grew bigger and stronger because they had reliable water, and economies benefited from healthier, more productive people.

"I will never be an old man. To me, old age is always 15 years older than I am." – Sir Francis Bacon

Sir Francis Bacon
Credit – ChatGPT/Image

KEY FIGURES IN WATER TREATMENT ADVANCEMENT

John Snow: Proved that contaminated water can cause disease, influencing public health policies.

James Simpson: Set up the first successful slow sand filtration system in London.

Dr. John L. Leal: Pioneered using chlorine for continuous water disinfection, helping eliminate waterborne illnesses.

EFFECT OF HYDRATION ON HEALTH AND LONGEVITY

It's not just *clean* water that matters—getting *enough* water every day is also crucial. Most experts suggest around 3.7 liters (about 125 ounces) per day for men and 2.7 liters (about 91 ounces) for women, though this includes liquids from foods like fruits and vegetables. Good hydration affects nearly every part of our bodies, from healthy digestion and brain function to helping our hearts work better. It can even reduce the risk of kidney stones and urinary tract infections.

Remember not to gulp too much water all at once—your body can only handle so much at a time. Spread your water intake throughout the day. Hydrate, but don't drown!

MEDIEVAL AND RENAISSANCE INVENTIONS

Let's look at some of the building blocks for our modern ideas about living longer:

St. Hildegard of Bingen (12th century): A Benedictine abbess who wrote about using natural remedies, stressing how a good diet and daily healthy habits can prevent illness.

Leonardo da Vinci: Studied human anatomy in astonishing detail. He believed that understanding how the body works could lead to better health and possibly extend life.

Leonardo Da Vinci
Source – Wikimedia Commons

Nostradamus: Best known for his prophecies, but he was also a doctor who cared for plague patients using practical, common-sense methods—such as promoting fresh air and cleanliness.

Sir Francis Bacon: Along with his interest in clean water, he believed that science and knowledge would lead to longer, healthier lives.

THE MODERN PURSUIT OF LONGEVITY
Scientific Advances and Psychological Insights
In the early 1900s, **Dr. Serge Voronoff** tried to slow down aging by grafting animal tissues into humans—controversial at the time (and discredited later), but it showed just how far some scientists would go to find the secret to living longer. Later research has shown that injecting older animals with younger blood can have limited rejuvenating effects, though it's still not a proven method for humans.

Carl Jung, a pioneer in psychology, felt that understanding our unconscious minds can lead to a healthier, more balanced life. Being mentally healthier can certainly contribute to a longer, more satisfying life.

Technological Innovations
Visionaries like **Nikola Tesla** and **Thomas Edison** revolutionized daily life with electricity and lighting, which improved living standards and opened the door for modern medical technology. **Albert Einstein's** theories eventually led to inventions such as MRI machines—key tools for detecting diseases early and treating them more effectively.

CONTEMPORARY VISIONARIES
Today, we see forward-thinkers in many fields pushing the boundaries of health, science, and technology. Here are just a couple of them:

STEVE JOBS, co-founder of Apple Inc., transformed technology into an accessible and integral part of daily life. His focus on innovation and design revolutionized our communication and access to information, indirectly promoting health education and awareness. **Jobs** was a once-in-a-century innovator who "saw into the future" of technology and changed the world as we know it. Although I'm not elaborating further here on **Jobs,** his contribution is extraordinary.

T. DENNY SANFORD
My dear friend **T. Denny Sanford** is a philanthropist who has donated significant sums to medical research, hoping to improve everyone's health and help people live better, longer lives. Through **Sanford Health** and other institutions that carry his name:

The Sanford Project: Focuses on curing type I diabetes.

Edith Sanford Breast Center: Advances breast cancer research and treatment.

Sanford Imagenetics: Looks at how our genes affect our health, aiming for more personalized care.

Sanford-Burnham Prebys Medical Discovery Institute: Researches everything from cancer to regenerative medicine.

Sanford Consortium for Regenerative Medicine: A group effort by top research centers to explore ways to repair damaged tissues and organs.

"Brains Into Space": Partly funded by Sanford, scientists have sent tiny lab-grown human brain cells into space to study how they function without gravity—another angle on understanding aging and disease.

T. Denny Sanford with Ron Zagami and Tad Sisler
Source – Sisler Private Collection

"People who are crazy enough to think they can change the world are the ones who do." – T. Denny Sanford

ELON MUSK

Elon Musk is known for Tesla (electric cars), SpaceX (space travel), and now **Neuralink**, which aims to connect our brains directly to computers:

How Neuralink Works

• It's a small device implanted in the skull with thin "threads" that read brain signals or even send signals back.
• A robotic system implants these threads to avoid damaging blood vessels.

Potential Health Benefits

• Helping paralyzed people move robotic limbs or communicate through "mind typing."
• Possibly treating Parkinson's disease, depression, and other conditions by detecting and modulating brain activity.
• In the long run, it might help slow or manage diseases like Alzheimer's, improving quality of life as people age.

Of course, there are always questions about ethics, safety, and privacy. But if it's developed responsibly, and I believe it will, **Neuralink** could open doors for people with serious medical problems—and maybe even the rest of us—by extending our healthy years.

"I think we have a duty to maintain the light of consciousness to make sure it continues into the future." — *Elon Musk*

Elon Musk
Credit — Wikimedia Commons

OTHER NOTABLE INDIVIDUALS CONTRIBUTING LARGE RESOURCES TO ADVANCE HEALTH AND LONGEVITY

BILL GATES

Although under extreme scrutiny from groups who believe that his entrepreneurship has been self-serving, and enduring claims that much of research he has funded and promoted is harmful to large groups of people, **Gates** has allocated billions of dollars through the *Bill & Melinda Gates Foundation* to combat infectious diseases, improve global health infrastructure, and fund medical research on vaccines and treatments.

NOTABLE CONTRIBUTIONS:

• Leading the drive to eradicate polio worldwide.
• Funding research and vaccine distribution for infectious diseases such as HIV/AIDS, malaria, and tuberculosis.
• Supporting advancements in gene editing and other cutting-edge medical technologies.

PATRICK SOON-SHIONG

A surgeon, medical researcher, and biotech entrepreneur, **Soon-Shiong** has invested heavily in finding innovative cancer treatments and other healthcare solutions.

NOTABLE CONTRIBUTIONS:

• Founder of *NantWorks,* focused on developing next-generation cancer treatments and personalized medicine.
• Pioneering efforts in immunotherapy and mRNA technology for cancer.
• Philanthropic support for hospitals and medical research, particularly in underserved communities.

JEFF BEZOS

Bezos has been increasingly directing resources toward longevity research and biotechnology.

NOTABLE CONTRIBUTIONS:

• Funding Altos Labs, a startup focused on cellular reprogramming to reverse disease and slow aging.

• Investing in various biotech ventures aimed at innovative health solutions.

• Supporting a broad range of science and technology initiatives, which may yield breakthroughs in health and disease prevention.

MARK ZUCKERBERG AND PRISCILLA CHAN

Through the *Chan Zuckerberg Initiative* (CZI), they are heavily invested in cutting-edge medical research, with the ambitious goal of helping to cure, prevent, or manage all diseases by the end of the century.

NOTABLE CONTRIBUTIONS:

• Funding the *Chan Zuckerberg Biohub*, which unites scientists and engineers to develop tools for diagnosing, treating, and preventing disease.

• Supporting large-scale initiatives in neuroscience, immunology, and artificial intelligence in healthcare.

LARRY ELLISON

The co-founder of *Oracle* established the *Ellison Medical Foundation* to explore the fundamentals of aging and age-related diseases.

NOTABLE CONTRIBUTIONS:

• Funding biomedical research focused on genetic, cellular, and organismal factors that influence lifespan.

• Supporting labs and universities that aim to better understand how to extend healthy human longevity.

Larry Ellison
Credit: Flickr/creativecommons.org

SPIRITUAL AND ALTERNATIVE PERSPECTIVES

Edgar Cayce, often called the "Sleeping Prophet," gave thousands of health and healing readings while in a trance. He believed in treating the whole person—body, mind, and spirit—and suggested natural remedies, healthy eating habits, and spiritual practices to support a long, healthy life. He stressed the importance of proper digestion and "keeping clean on the inside" through regular bowel movements—something that still makes sense today!

We can already see how these spiritual or alternative ideas tie in with what we know about modern longevity research. Through history, people have tried all kinds of approaches—mythical, spiritual, scientific—to figure out how to live longer. Each era brings its own unique spin on the question, but they all share the same goal: to make life not only longer, but better.

"We are not victims of aging, sickness and death. These are part of scenery, not the seer, who is immune to any form of change. The seer is the spirit, the expression of eternal being." – Deepak Chopra

As we stand on the edge of new discoveries in biotechnology and artificial intelligence, these older perspectives remind us that living longer isn't just about adding more years—it's about enjoying those years with a good quality of life. This search for longevity has always been about hope and curiosity—two things that have driven human progress from the start.

HISTORY OF AGING AND LIFE EXPECTANCY

HUMAN LIFESPAN EVOLUTION
PRESHISTORIC TIMES TO PRESENT

When I was growing up and studying the Bible, I learned that one of God's first commands was, *"Be fruitful and multiply."* Since ancient times, people have strived to survive and thrive under tough conditions. In prehistoric days, the average person only lived to around 30 years. Harsh environments, lack of reliable food, and constant threats from predators or diseases all played a part. Things started to change about 10,000 B.C. with the arrival of agriculture. Living in one place meant more steady food supplies, but it also brought new problems, like contagious diseases spreading through crowded settlements. Over time, even though life remained challenging, stable farms did help inch average lifespans upward.

In the Middle Ages, progress was slow. Wars, famine, and plagues (like the Black Death) often wiped out whole populations. Big improvements didn't really show up until the Enlightenment and the Industrial Revolution.

Technology, hygiene, and better public health measures all started to make a difference. For example, in late-1800s New York City, horse manure piling up on the streets created horrible smells and disease. Once people realized they had to clean it up, infection rates dropped. Little by little, these changes helped people live longer.

IMPACT OF AGRICULTURE, INDUSTRIALIZATION, AND MODERN MEDICINE ON LIFE EXPECTANCY

The Agricultural Revolution gave us more reliable food, but living so close together also led to faster spread of diseases. Many people were malnourished, and doctors didn't yet understand how diseases really worked.

Later on, during the 18th and 19th centuries, industrialization and urbanization led to overcrowded cities with poor sanitation. That made illness spread like wildfire—at first. Over time, though, these crowded conditions pushed city leaders to develop sewer systems, cleaner water, and better waste management. Those improvements helped slash infection rates.

The 20th century brought huge medical breakthroughs—especially antibiotics and vaccines. My great-grandfather was a Union soldier in the Civil War, and his family treated his wounds with a bark poultice that acted like a natural antibiotic. They didn't even realize that's what it was! Years later, **Alexander Fleming's** discovery of penicillin in 1928 changed medicine forever. During World War II, my father saw how precious penicillin could be on Navy ships. In extreme cases, if medicine was running low, doctors sometimes tried desperate measures—like having other sailors drink the urine of the patient who got the penicillin—to keep it circulating. It sounds shocking, but during war, it saved lives.

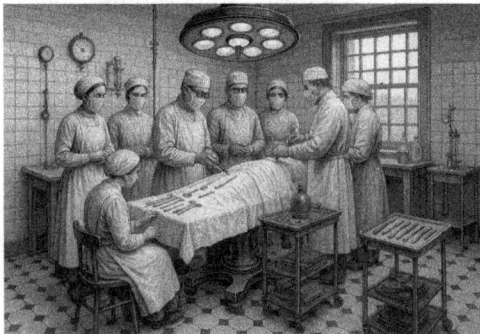

Vaccines also began to wipe out diseases that once seemed unbeatable, like smallpox and polio. By the end of the 20th century, better surgery, nutrition, and education helped push average life expectancy from the mid-40s into the 70s in many developed countries. Today, there's growing debate about too many vaccines too soon, but we can't ignore how many lives they've saved.

MODERN SCIENTIFIC REVOLUTION – BREAKTHROUGHS IN GENETICS, BIOTECHNOLOGY, AND MEDICINE

Decoding the human genome has opened the door to understanding what makes us age. Scientists have found certain genes linked to longevity and are experimenting with advanced tools like CRISPR-Cas9 to fix or remove harmful genes. Companies working in biotech are finding ways to slow cellular aging and keep cells healthier for a longer time. They're also exploring how to regrow or replace damaged tissues, like cartilage or even whole organs, using stem cells.

We'll talk more about these amazing developments in the chapters to come. My goal with this book is to give you enough science to understand the major breakthroughs—without overwhelming you. I hope you'll find that, even if some concepts seem complicated at first, they have the potential to change our lives in ways we could never have imagined. And that, to me, is what keeps the journey exciting.

PROLOGUE
CURRENT ENVIRONMENTAL CONCERNS

The Industrial Revolution transformed our world in both good and bad ways. While many inventions from this era have raised our chances for living longer, they've also introduced new diseases linked to chemical exposure—like mesothelioma, silicosis, and Minamata disease. Even now, we worry about the effects of industrial byproducts on our health. Before we dive into modern science and medicine, let's look at some outside factors that might keep us from living longer, healthier lives.

ELECTRICAL FREQUENCIES- ENVIRONMENTAL RADIATION

Do everyday electrical frequencies and radiation hurt us, or help us? We still don't have a final answer, but research is ongoing. People often worry about microwaves, cell phone signals, or 5G radiofrequency. Here's what we know so far.

ELECTRICAL FREQUENCIES (Extremely Low Frequency, ELF)

Where It's Found: Power lines, household wiring, and appliances (around 50–60 Hz).

Potential Issues: Some studies hint at a small link between high ELF exposure (like living right under a power line) and slightly higher rates of childhood leukemia. But no firm cause-and-effect has been proven, and the overall risk appears very low.

Possible Benefits: In controlled medical settings, pulsed electromagnetic field (PEMF) therapy can help heal bones or tissues. If you've ever had muscle stimulation during physical therapy, that's a form of safe ELF use.

RADIOFREQUENCY (RF) FIELDS (Wi-Fi, Cell Phones, 5G)

Where It's Found: Frequencies from hundreds of MHz to tens of GHz. 5G can range from about 600 MHz to around 39 GHz.

Potential Concerns: After decades of research, we haven't found solid proof that normal day-to-day RF exposure (like from your phone or Wi-Fi) causes cancer. However, some people report headaches, anxiety, or insomnia they believe come from RF exposure. Scientists often attribute these to the "nocebo" effect—where we worry ourselves into feeling ill.

Potential Positives: 5G makes telemedicine easier, bringing fast healthcare services to remote areas. RF is also used in MRI machines, which have revolutionized the way we diagnose diseases.

IONIZING VS. NON-IONIZING RADIATION

Ionizing Radiation: Includes X-rays and gamma rays, which can damage DNA and raise cancer risk. Still, controlled medical use (like radiation therapy) saves lives.

Non-Ionizing Radiation: Includes visible light, microwaves, and radio waves. These don't usually break down DNA; their main effect at high levels is heating tissue.

Bottom Line: Everyday exposures to non-ionizing radiation (like 5G) are generally considered safe if they follow safety rules. Ionizing radiation can be harmful in large amounts, but in medicine, it can also detect and treat serious conditions.

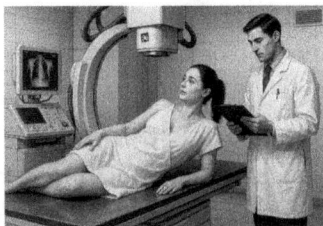

THREAT OF ENVIRONMENTAL TOXINS TO HUMAN HEALTH

In the early 1970s, I remember visiting my wonderful grandmother, **Gizella Witt**, in Long Beach, California. The Los Angeles Basin regularly had a yellowish color from the smog. My uncle took me to Guadalajara, Mexico, around the same time, and the air was so filthy it was brown.

Although our air has been purified quite a bit since those times due to the use of catalytic converters on automobiles and regulations curtailing the pouring of smoke into the air, some Worldwide cities still reel from industrial waste.

There's still a big concern about chemicals and pollution. Some cities continue to suffer from smog and industrial waste. For example, near where I live in San Diego, the Tijuana River dumps human and industrial waste into the Pacific Ocean daily, making the water unsafe for swimmers in Imperial Beach.

THE EVOLVING LANDSCAPE OF CHEMICALS

Legacy Contaminants: Things like asbestos, lead paint, and DDT are banned in many countries, but they linger in older buildings, soil, and water.

Persistent Organic Pollutants (POPs): PCBs and dioxins break down extremely slowly and can cause hormone disruption and other issues.

Emerging Concerns: PFAS ("forever chemicals") used in non-stick pans and flame retardants can accumulate in our bodies and may be linked to problems like metabolic disorders, reproductive issues, and cancer.

WHAT CAN WE DO?

Stricter Standards & Cleanups: Governments can push for better regulations and enforce existing ones more strictly. Cleaning up polluted sites reduces harmful exposure.

Workplace Safety: We need good industrial hygiene, protective gear, and thorough training, so workers aren't endangered by toxic chemicals.

Consumer Awareness: Read labels, choose products with fewer harsh chemicals, and keep your home well-ventilated. Detectors for carbon monoxide or smoke are essential.

Healthy Lifestyle: A good diet, exercise, and regular check-ups help your body stay strong against low-level toxins that can add up over time.

SUPPLEMENTS OR MEDICINES FOR TOXIN EXPOSURE

No supplement or medication can guarantee complete protection against toxins or cancer, and research often shows mixed results. Here are some nutrients that get studied:

Antioxidants (Vitamins C & E): Can neutralize free radicals caused by certain toxins, but large trials haven't proven they stop cancer.

Selenium: Early studies looked promising, but later trials didn't confirm a strong protective effect.

Vitamin D: Linked to better health in general, but not a magic shield against environmental toxins.

Green Tea Polyphenols, Cruciferous Veggies, Curcumin (Turmeric): Some lab studies suggest these may help the body handle stress from toxins, but human evidence is still limited.

For more information, I go into extensive detail on every important vitamin and supplement in my book **Vitamins, Supplements, and Herbs for Health and Longevity: Boost Your Immunity, Increase Energy, and Feel Younger in Minutes a Day.**

PRACTICAL TIPS

Whole Foods First: Eat a balanced, plant-rich diet.

Lifestyle Habits: Don't smoke, limit alcohol, maintain a healthy weight, and stay active.

Seek Professional Advice: Always consult a healthcare provider before starting any supplement routine. There is great wisdom in the words of my friend, legendary actress **Mary Tyler Moore:**

"You truly have to make the very best of what you've got. We all do."

Tad Sisler with Mary Tyler Moore
Source- Sisler Private Collection

In short, there's no single pill that cancels out chemical exposure. But living a healthy, balanced life helps reduce the risks.

DO ALL BODY CELLS REPLACE THEMSELVES?

You might've heard that every cell in your body replaces itself every seven years. That's not entirely true. Cell turnover varies:

Fast Turnover: Gut and blood cells renew in days or weeks. Skin replaces itself in about a month.

Moderate Turnover: Liver cells might renew every few hundred days.

Slow or No Turnover: Neurons in your brain's cortex mostly stick around for life. Same for the lens in your eye and heart muscle cells.

Your body is a patchwork of cells with different lifespans, and there's no full "reset" where everything is replaced at once.

CAN UNHEALTHY LIVING BE REVERSED?

Some damage is permanent, but there's good news: many healthy habits can halt or even partially reverse harm done by poor lifestyle choices.

Weight Loss & Diet: Dropping excess weight helps control blood sugar, blood pressure, and reduces inflammation.

Quit Smoking & Limit Alcohol: Stopping smoking can dramatically lower the risk of lung cancer and heart disease. Cutting back on alcohol helps protect the liver and reduces certain cancer risks.

Exercise & Sleep: Staying active boosts heart health, muscle mass, and helps manage stress. Good sleep and stress reduction can fight chronic inflammation and support healthy aging.

Mental Health: A positive outlook and good coping strategies can lower stress hormones that speed up aging.

Adopting stress-management techniques (e.g., meditation and yoga) and improving sleep hygiene can reverse some of these negative impacts over time and enhance overall health span. My friend, legendary rapper **Snoop Dogg,** had his perspective:

"You've got to always go back in time if you want to move forward."

Snoop Dogg and Tad Sisler
Credit – Sisler Private Collection

New breakthroughs in science hopefully will allow us to reverse the aging process soon, but meanwhile, living healthy is a solid bet.

STAY OUT OF TROUBLE!

Even your environment—like being in prison—can affect how fast you age. Chronic stress, inadequate healthcare, poor diet, and lack of sunlight can cause prisoners to develop diseases and frailty sooner. Not everyone in prison experiences this at the same rate, but it shows how stress and isolation can take a serious toll on the body.

My book **"The Science of Positive Thinking"** illustrates that mindset, daily habits, and emotional well-being can add years to one's life, no matter where one finds themselves.

SCREENING HEAVY METALS WITH HAIR ANALYSIS

Hair analysis can reveal some heavy metals (like mercury or lead) that you've been exposed to over the past few months. It's not a perfect test—hair products can skew results, and different labs use different methods. If hair tests show something suspicious, doctors usually follow up with blood or urine tests to confirm.

WHY CHECK HEAVY METALS?

Oxidative Stress & Inflammation: Metals like lead or cadmium can hurt cells and raise the risk of chronic diseases.

Brain & Organ Health: High exposures can damage nerves, kidneys, bones, and more.

Cancer Risk: Some metals increase cancer risk over time.

MINIMIZING YOUR EXPOSURE

Water Quality: Test your water, especially if you live in an older home with lead pipes.

Seafood Choices: Eat smaller fish like salmon or sardines rather than big ones like tuna or swordfish, which can contain more mercury.

Air & Household Products: Keep indoor air fresh and choose products with fewer harsh chemicals.

Professional Checkups: See a doctor for testing if you suspect high exposure.

THE ROLE OF PERSONAL HYGIENE

Staying clean isn't just about smelling good—it's one of your best defenses against germs. Regular handwashing, bathing, and oral care help prevent infections that wear your body down over time.

Lower Infection Risk: Fewer illnesses mean less inflammation and stress on your body.

Skin Protection: Your skin is your largest barrier. Keeping it clean and moisturized helps it function its best.

Oral Health: Good brushing and flossing can protect your heart and body from chronic inflammation.

Mental Well-Being: Feeling clean boosts confidence and lowers stress, which may have long-term health benefits.

GEROSCIENCE: STUDYING AGING ITSELF

Geroscience looks at how aging causes age-related diseases like Alzheimer's or heart disease. By understanding aging at the molecular level—things like telomere shortening and mitochondrial breakdown—scientists hope to slow or prevent these conditions.

LEADING RESEARCHERS

Dr. David Sinclair (Harvard): Studies sirtuins and molecules like resveratrol (found in red wine) that may mimic calorie restriction and support longer lifespans.

"Only 20 percent of our longevity is genetically determined. The rest is what we do, how we live our lives and increasingly the molecules that we take. It's not the loss of our DNA that causes aging, it's the problems in reading the information, the epigenetic noise." – Dr. David Sinclair

Dr. Cynthia Kenyon: Found a gene in tiny C. elegans worms that can double their lifespan, linked to insulin and growth factors in humans.

"Age is the single largest risk factor for an enormous number of diseases. So, if you can essentially postpone aging, then you can have beneficial effects on a whole range of disease." – Dr. Cynthia Kenyon

LIFESPAN VS. HEALTHSPAN

Lifespan is how long you live; **healthspan** is how long you stay healthy and active. Modern research focuses more on healthspan because who wants to live extra years in poor health?

BREAKTHROUGH BOOKS IN LONGEVITY

Lifespan: Why We Age—and Why We Don't Have To (Dr. David Sinclair) Argues that aging is a disease we can fight. Introduces the "Information Theory of Aging," suggesting we lose "cellular information" as we get older.

Life Force (Tony Robbins, Peter Diamandis, Robert Hariri)
Explores regenerative medicine, gene editing, stem cells, and how to take charge of your own health.

Reading **Tony Robbins'** *Life Force* was a turning point in my life. I had no idea that so many breakthroughs in regenerative medicine and personalized health were happening at once. Co-authored with **Peter H. Diamandis** and **Robert Hariri**, the book delves into cutting-edge technologies like stem cell therapy, gene editing, and synthetic biology. It emphasizes taking proactive control of one's health through traditional and innovative approaches. These books get us thinking about not just living longer, but living better. They also push us to consider the ethics of extending human life.

> *"There's no such thing as failure. There are only results."*
> *— Tony Robbins*

Tony Robbins
Credit – Wikimedia Commons

LESSONS FROM ANIMALS

Certain animals—like bowhead whales (which can live 200 years) or naked mole rats—age very slowly or resist cancer. Studying their genes and body chemistry could teach us secrets to healthier aging. Some researchers also work on extending the lifespans of dogs, which could eventually help humans. Personally, I'd love to see my Pomeranian, Frankie, live much longer!

BECOMING EMPOWERED

This book shares the latest findings in genetics, epigenetics, and biotech—plus everyday tips on nutrition, exercise, and stress management. While you should always consult a doctor for personal medical questions, I hope to show you realistic ways to boost your healthspan. If we can stay open-minded, there's no telling how far we can push the limits of aging. Let's begin our journey together to see how long we can keep our quality of life. How wonderful it would be to see our great-grandchildren grow up, to see new advances in technology, and maybe we'll still be around too to enjoy them! If we keep an open mind, we will allow ourselves to discover new possibilities we may not have otherwise allowed. My old friend, **President George H.W. Bush** described open mindedness in this fashion:

"I have opinions of my own, strong opinions, but I don't always agree with them."

President George H.W. Bush and Barbara Bush with Tad Sisler
Source – Sisler Private Collection

PART I
UNDERSTANDING THE BIOLOGY OF AGING

CHAPTER ONE
THE HALLMARKS OF AGING

My lovely grandmother **Audrey Athey Sisler** was born in 1889. In her lifetime, she witnessed humankind going from horse and buggy to exploring the moon. She was already almost seventy when I was born, but she had enough life and fervor left in her to become my best friend, living to see my twins before she passed in 1983 at 94.

Tad Sisler's Grandmother, Audrey Athey Sisler
Source – Sisler Private Collection

My grandmother **Audrey** told me, *"I swear, I'm still eighteen inside!"* and I believed her. The first and easiest thing we can do is to understand that our attitudes and emotions play a huge role in success in life. I believe strongly that an optimistic outlook adds to our life expectancy as well. Even through all of the tragedy she experienced, my grandmother taught me optimism and so much more. My friend, *Academy-Award-winning* actress **Cloris Leachman** said:

"I don't think I'm my age. I'm truly 6 years old."

Academy-Award-Winning Actress Cloris Leachman and Tad Sisler
Source- Sisler Private Collection

Well, that's a stretch, but you get the idea. Remaining childlike and young at heart is most likely a key to longevity.

My greatest teachers were those who were able to explain things in terms I could understand, whether through analogies, storytelling, or metaphors. I will attempt to do this throughout the book, first presenting ideas in sophisticated terms, and then explaining it as best as I can in a way that anyone may be able to understand.

OVERVIEW OF THE HALLMARKS OF AGING

Aging is like a puzzle with many pieces, and scientists have identified nine key factors—called hallmarks—that explain why our bodies change as we get older. These hallmarks are the building blocks of aging, and understanding them helps us find ways to stay healthier for longer. Here's a quick look at all nine, with details on some in this chapter and others later in the book:

Genomic Instability: Damage to our DNA that builds up over time, like smudges in a recipe book (covered in this chapter).

Telomere Attrition: Shortening of the protective caps on our chromosomes, like fraying shoelace tips (covered in this chapter).

Epigenetic Alterations: Changes in how our genes are turned on or off, like adjusting a playlist's volume (covered in this chapter).

Loss of Proteostasis: Problems with protein folding and cleanup, like tangled earbuds causing trouble (covered in this chapter).

Deregulated Nutrient Sensing: When our body doesn't respond well to food and energy, like a faulty gas gauge (see Chapter 3).

Mitochondrial Dysfunction: When our cell powerhouses start to fail, like a car engine losing steam (see Chapter 3).

Cellular Senescence: Cells that stop dividing and cause inflammation, like lazy workers slowing things down (see Chapter 2).

Stem Cell Exhaustion: Fewer new cells to repair our body, like running out of spare parts (see later chapters).

Altered Intercellular Communication: Mixed-up signals between cells, like a bad phone connection (see later chapters).

This chapter dives deep into the first four hallmarks, giving you a solid start on understanding aging and what you can do about it right now.

GENE STUDIES
GENOMIC INSTABILITY
DNA DAMAGE AND REPAIR MECHANISMS

It amazes me how far we've come since cracking the genome around the turn of the 21st century. Our understanding of DNA and RNA is still in infancy, but we have learned much already. Exponential advances in AI and computational capabilities increase our understanding of genetics, with new developments coming almost daily. Imagine when this knowledge multiplies by a billionfold, as **Ray Kurzweil** prophesies!

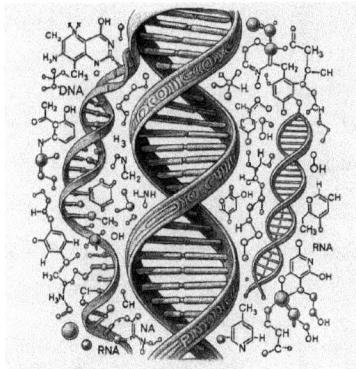

DNA

Genomic instability is a fundamental hallmark of aging, characterized by an increased frequency of mutations within the genome due to DNA damage. This damage arises from endogenous sources like reactive oxygen species generated during metabolic processes and exogenous sources such as ultraviolet (UV) radiation, ionizing radiation, and environmental mutagens. DNA damage can manifest as single-strand breaks, double-strand breaks, crosslinks, and base modifications.

Think of your DNA as a giant instruction manual for your body, telling your cells how to grow, divide, and do their jobs. Sometimes, mistakes happen in the instructions because of too much sun (radiation) or chemicals (oxidative stress). It's like getting smudges or tears in a book.

Cells have evolved intricate DNA repair mechanisms to maintain genomic integrity:

Nucleotide Excision Repair (NER): Repairs bulky helix-distorting lesions, such as those caused by UV light-induced thymine dimers.

Base Excision Repair (BER): Corrects small, non-helix-distorting base lesions resulting from oxidation, deamination, and alkylation. Imagine cleaners (**NER** and **BER**) who erase smudges and fix small tears.

Mismatch Repair (MMR): Fixes replication errors like misincorporated bases and insertion-deletion loops.

Homologous Recombination (HR) and Non-Homologous End Joining (NHEJ): Repair double-strand breaks through error-free and error-prone pathways, respectively. Think of maintenance specialists who fix big rips in the pages.

Dr. Jan Vijg's research emphasizes the role of genome maintenance in aging. He suggests that the accumulation of DNA damage and mutations impairs cellular function and accelerates aging (*Vijg, J. "Aging and genome maintenance." Mechanisms of Ageing and Development, 2014*).

Recent Development: In February 2025, a study in *bioRxiv* reported a new small molecule that enhances DNA repair by boosting the activity of BER pathways in aged mice, reducing genomic instability and improving healthspan. This suggests potential therapies to protect DNA as we age, aligning with the book's focus on cutting-edge solutions.

WHY IT MATTERS
Just like a book with missing pages or blurry words can't tell a story properly, cells with too much DNA damage can't function well. This leads to aging and diseases that often happen as people age.

MUTATIONS AND THEIR ACCUMULATION THROUGH TIME
What are they? Somatic (body cell) mutations happen throughout an organism's life. These are small changes in the DNA of cells that aren't passed on to offspring but build up over time in our own tissues.

Why do they matter? When too many mutations accumulate, they can disrupt normal cell function, causing cells to stop dividing (senescence), self-destruct (apoptosis), or become cancerous.

How are they related to aging and disease? This ongoing accumulation of genetic changes is linked to many age-related illnesses, including cancer, certain brain disorders, and heart diseases.

TELOMERE ATTRITION
Role of Telomeres in Cellular Aging
What are telomeres? Telomeres are repetitive DNA sequences at the ends of chromosomes. They act like protective caps, preventing the ends of our chromosomes from fraying or fusing with each other.

Why do they shorten? Each time a cell divides, its telomeres get a bit shorter because DNA-copying enzymes can't fully replicate the very ends of chromosomes.

What happens when they get too short? Critically short telomeres cause cells to see this as DNA damage, stopping cells from dividing or triggering cell death. This process is an important factor in aging and in conditions that worsen with age.

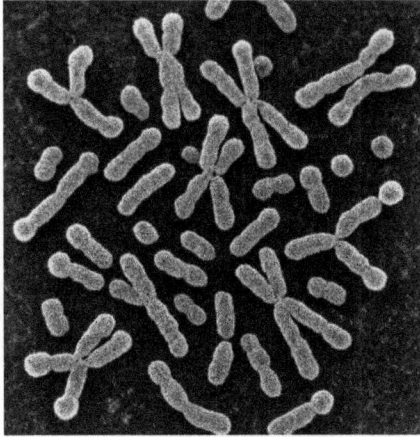

Telomeres

TELOMERE ACTIVATION STRATEGIES

How can telomeres be lengthened? Telomerase is an enzyme that can extend telomeres. In most adult cells, telomerase is turned off. In some cells—like germ (egg and sperm) cells—telomerase is active, and these cells maintain longer telomeres.

Possible interventions:

Gene therapy: Introducing or increasing the activity of telomerase in cells.

Drugs (like TA-65): Small molecules that claim to enhance telomerase activity.

Risks: If cells keep dividing without limit (thanks to active telomerase), there's a higher risk of cancer because these cells can potentially grow unchecked.

Overall, the takeaway is that mutations accumulating over time and telomere shortening both play central roles in how cells age, contributing to various diseases. While boosting telomerase activity could help cells stay "younger," it also raises the concern of promoting uncontrolled cell growth.

Recent Development: In March 2025, researchers at Stanford University developed a non-invasive blood test to measure telomere length more accurately, allowing doctors to track biological aging and tailor interventions like exercise or stress management to protect telomeres (Telomere Measurement).

Think of telomeres as the plastic tips at the ends of your shoelaces (called aglets). They keep your shoelaces from fraying. Every time your cells divide, the telomeres get a little shorter, like if the plastic tips wore down each time you tied your shoes. Eventually, they become too short, and the shoelaces start to fray, which makes the cells stop working correctly. Telomerase is like a magic glue that can rebuild the plastic tips, making the shoelaces (telomeres) longer again.

Scientists are exploring ways to use telomerase to keep telomeres from getting too short, hoping this could slow down aging. But they must be careful because using too much "magic glue" might cause problems, like cells growing out of control (which can lead to cancer).

WHY IT MATTERS

Keeping telomeres long might help cells stay healthy longer, like keeping shoelaces from fraying so your shoes last longer. But it's important to find a balance to avoid unwanted side effects.

"What I found out on Christmas Day 1984, through biochemical evidence, was that telomeres could be lengthened by the enzyme we call telomerase, which keeps the telomeres from wearing down. After I found that out, I went home and put on Bruce Springsteen's 'Born in the USA' which was just out, and I danced and danced and danced." – Carol W. Greider

Carol Greider
Credit – Wikimedia Commons

EPIGENETIC ALTERATIONS

Think of your DNA as a big instruction manual. Epigenetics refers to small "tags" or "markers" added onto this manual (or the proteins that package it), which change how you read the instructions without changing the words themselves (the DNA sequence).

These modifications act like dimmer switches for genes, turning them up or down, rather than rewriting their code.

TYPES OF EPIGENETIC MODIFICATIONS

DNA Methylation: This is like putting a chemical tag (a methyl group) on certain letters of your DNA. Usually, more tags mean the gene is turned down or off.

Histone Modifications: DNA is wrapped around proteins called histones. Adding or removing chemical marks on histones changes how tightly DNA is wound, which affects whether genes are "open" (easier to read) or "closed" (harder to read).

Non-coding RNAs: These are RNAs that don't directly make proteins but help control which genes are active.

AGING AND DYSREGULATION

As we get older, these epigenetic tags often change in unhelpful ways, leading to genes being turned on or off at the wrong times. This can cause cells to function poorly and throw off the body's balance.

Dr. Steve Horvath developed an "epigenetic clock" that measures how these DNA methylation patterns shift with age. It can predict someone's "biological age," which may differ from their actual chronological age.

REVERSIBLE VS. IRREVERSIBLE CHANGES

The good news: Epigenetic changes can sometimes be reversed. Scientists can use techniques (like induced pluripotent stem cell technology) to "reset" these tags, making cells behave as if they are younger. This is one of the more exciting developments in age reversal research.

The challenge: Fully resetting all tags at once can be dangerous. It might accidentally push cells to lose their normal identity or even become cancerous. So, any therapy that tries to reverse aging through epigenetics must be done carefully to avoid these risks.

Recent Development: In April 2025, a clinical trial reported that combining meditation with a polyphenol-rich diet significantly reduced epigenetic age in participants, as measured by Horvath's clock, offering a natural way to slow aging (Epigenetic Reversal).

In short, epigenetics is about HOW our body reads our DNA. Over time, these reading patterns can shift, leading to signs of aging. The hope lies in finding safe ways to correct or reset these patterns so cells can work better and possibly even slow down or reverse some aspects of aging.

Think of your genes like a music playlist. Epigenetics is like each song's volume controls and on/off switches. As you get older, some songs that should be playing get turned down or off, and others that shouldn't be playing get turned up.

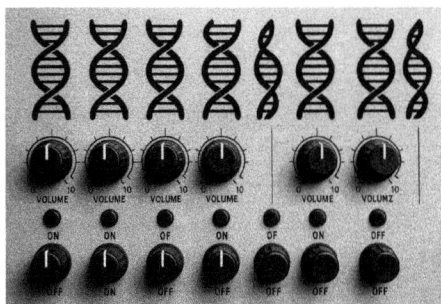

Dr. Steve Horvath's "clock" can tell how old your cells are by looking at these volume settings (DNA methylation patterns). Scientists are looking for ways to adjust the volume back to the settings from when you were younger, hoping to keep cells functioning well. If we can figure out how to reset these switches, we might be able to "re-tune" our cells to stay healthier as we age, like updating a playlist to keep it fresh and enjoyable.

LOSS OF PROTEOSTASIS (PROTEIN BALANCE)

Protein Folding and Aggregation: In medical terms, proteostasis involves regulating the cellular protein pool through synthesis, folding, trafficking, and degradation. Molecular chaperones assist in proper protein folding. Misfolded proteins can aggregate, forming toxic oligomers and fibrils that disrupt cellular function.

Cells carefully balance the making, folding, moving, and breaking down of proteins (this balance is called proteostasis). *Molecular chaperones* are special proteins that help other proteins fold correctly. If proteins don't fold correctly, they can form harmful clumps (aggregates). These clumps can damage cells and lead to diseases like:

Alzheimer's Disease: Linked to β-amyloid plaques and tau tangles.

Parkinson's Disease: Linked to α-synuclein clumps called Lewy bodies.

AUTOPHAGY (CELLULAR CLEANUP)

Autophagy is the cell's recycling system: it breaks down and reuses worn-out cell parts and misfolded proteins in structures called lysosomes. There are different types, like macroautophagy, microautophagy, and chaperone-mediated autophagy. Boosting autophagy can help cells deal with harmful protein buildup and function better.

WAYS TO INCREASE AUTOPHAGY:

Caloric Restriction: Eating fewer calories (without causing malnutrition) has been shown to spur autophagy.

Drugs: Medications like rapamycin and metformin can imitate some effects of caloric restriction and may also enhance autophagy.

Recent Development: In May 2025, a new autophagy-inducing drug entered phase II clinical trials, showing promise in clearing protein aggregates linked to Alzheimer's disease, potentially offering a new way to maintain proteostasis (Autophagy Drug Trial).

Proteins in your cells must be folded just right to work, like origami shapes. Sometimes, they fold the wrong way and stick together, forming clumps like tangled earbuds. These clumps can cause brain problems like Alzheimer's disease. Autophagy is like your cell's recycling and trash service. It cleans up broken parts and misfolded proteins. You can boost this cleanup crew by eating less (caloric restriction) or taking certain medicines.

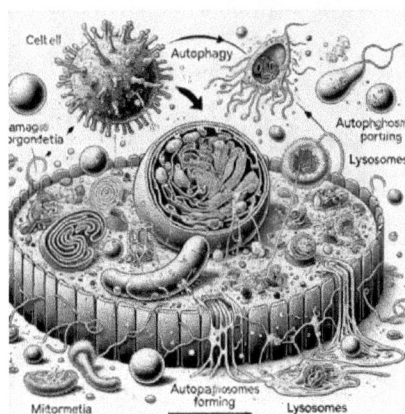

Properly folding proteins and cleaning up the misfolded ones helps cells stay healthy. It's like putting away shoes so you don't trip over them.

PULLING IT ALL TOGETHER

Understanding these hallmarks of aging provides insights into the complex biological processes that lead to age-related decline. Exploring genomic instability, telomere attrition, epigenetic alterations, and loss of proteostasis, will help researchers to develop interventions that could possibly extend our lifespan and improve health during aging.

Scientists are like detectives, studying why our bodies change as we age. They're looking at how our DNA gets damaged, how the protective ends of our chromosomes (like shoelace tips) wear down, how the "switches" that control our genes get messed up, and how proteins in our cells sometimes fold the wrong way and cause trouble.

By understanding these changes, they hope to find ways to help us stay healthier and maybe even live longer—possibly up to 150 years! It's like figuring out how to keep a car running smoothly for many more miles by caring for the engine, fixing any problems, and using the best fuel.

"The body itself is an information processor. Memory resides not just in brains but in every cell. No wonder genetics bloomed along with information theory. DNA is the quintessential information molecule, the most advanced message processor at the cellular level — an alphabet and a code, 6 billion bits to form a human being." — James Gleick

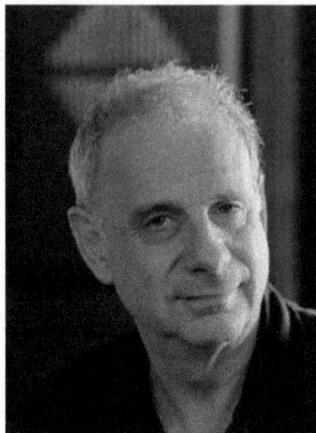

James Gleick
Credit — Wikimedia Commons

PROTECTING YOUR DNA AND GENOMIC STABILITY WHAT YOU CAN DO NOW

STAY SAFE FROM EXCESSIVE RADIATION

Sun Protection: Wear hats, sunglasses, and low- or no-chemical sunscreen, to shield your skin from the sun's harmful rays.

Limit Unnecessary X-Rays: Only have medical scans like X-rays or CT scans when your doctor says it's necessary.

EAT ANTIOXIDANT-RICH FOODS

Colorful Fruits and Veggies: Foods like berries, oranges, spinach, and broccoli help protect your cells from damage.

MAINTAIN HEALTHY HABITS

Don't Smoke: Smoking can harm your DNA and speed up aging.

Get Enough Sleep to help your body repair itself, which includes fixing damaged DNA.

EXCITING THINGS COMING SOON
BETTER DNA REPAIR

Scientific Research: Scientists like **Dr. Jan Vijg** are studying how our bodies fix DNA to keep cells healthy.

Future Treatments: In the coming years, new medicines might help our cells repair DNA better, slowing down aging and preventing diseases.

46

PROTECTING CHROMOSOME CAPS: TELOMERE ATTRITION
WHAT YOU CAN DO NOW
STAY ACTIVE

Regular Exercise, like running, swimming, or playing sports, helps keep cells young by protecting telomeres (the caps on chromosomes).

MANAGE STRESS

Relax and Unwind: Doing things you enjoy, like reading, drawing, playing music, or spending time with friends, can reduce stress and protect your telomeres.

EAT A HEALTHY DIET

Balanced Meals: Eating lots of fruits, vegetables, and whole grains supports overall cell health.

We all tend to forget that our bodies need regular preventative maintenance. If we treated our bodies like our cars, we would all be in better shape. **Steve Jobs** had a great take on this idea; nurture your body and mind although your actions may never be seen by others, but you know what you've done; you've taken the right steps to make yourself as whole and healthy as possible:

"When you're a carpenter making a beautiful chest of drawers, you're not going to use a piece of plywood on the back, even though it faces the wall and nobody will ever see it. You'll know it's there, so you're going to use a beautiful piece of wood on the back." – Steve Jobs

Do only the best and healthiest things you can do for your body, your temple. If you have a 'cheat' day, it's ok; just pick yourself up, dust yourself off, and start all over again. You'll feel better about yourself, I promise. For a greater grasp of weight management and proper diet, check out my book **The Ultimate AI Diet - Consolidating the Best Diets Over the Last 100 Years.**

EXCITING THINGS COMING SOON
TELOMERASE RESEARCH

Studying Cell Enzymes: Scientists are continuing to explore telomerase, an enzyme that can lengthen telomeres.

Potential Therapies: In the future, treatments might help keep telomeres longer, possibly slowing aging—but scientists need to make sure they're safe and won't cause problems like cancer.

ADJUSTING GENE ACTIVITY: EPIGENETIC ALTERATIONS
WHAT YOU CAN DO NOW
HEALTHY LIVING

Avoid Harmful Chemicals: Stay away from pollutants and toxic substances that can change how your genes work.

NOURISH YOUR BODY

Foods Rich in B Vitamins: Eating leafy greens, whole grains, and eggs can support healthy gene activity.

STAY ACTIVE AND STRESS-FREE

Exercise Regularly: Physical activity can have positive effects on your genes.

Relaxation: Activities like yoga, meditation, or simply spending time outdoors can help. Learn to "reframe" stress.

"Enjoy the pressure. Enjoy the stress. Enjoy being uncomfortable. And don't shy away from it, embrace it." – Gary Woodland

NEW DEVELOPMENTS IN RESETTING THE AGING CLOCK

Epigenetic Research: Scientists like **Dr. Steve Horvath** are figuring out how changes in gene activity affect aging.

Future Possibilities: Maybe one day, we'll be able to "reprogram" our cells to act younger, improving health and longevity.

KEEPING PROTEINS HEALTHY: LOSS OF PROTEOSTASIS
WHAT YOU CAN DO NOW
EAT FOR PROTEIN HEALTH

Antioxidant Foods: Fruits like blueberries and vegetables like kale help protect proteins in your body.

STAY ACTIVE

Regular Exercise: Helps your body manage proteins better and keeps cells functioning properly.

BOOST CELLULAR CLEANUP WITH INTERMITTENT FASTING

With Medical Guidance: Sometimes, short periods of fasting (skipping a meal) can help your body's cleanup processes—but only do this if it's safe and a medical professional approves.

EXCITING THINGS COMING SOON
IMPROVING AUTOPHAGY

Scientific Discoveries: Researchers are finding ways to enhance autophagy, the process where cells clean out damaged parts.

Disease Prevention: In the future, boosting autophagy might help prevent diseases like Alzheimer's by keeping proteins from clumping together.

PULLING IT ALL TOGETHER

While scientists are working hard on new ways to help us live longer and healthier lives, there are many things you can do right now:

Live a Healthy Lifestyle: Eat nutritious foods, stay active, get enough sleep, and avoid harmful substances.

Protect Yourself: Use sunscreen, wear protective clothing, and stay away from pollutants.

Manage Stress through activities that allow you to relax and enjoy life.

LOOKING AHEAD

Stay Curious: Keep learning about new scientific discoveries that could help us age better. As we delve deeper into our body on a cellular level, new medical and surgical techniques will naturally be invented that will solve many problems regarding keeping our cells youthful and strong.

Future Treatments: Exciting new therapies might be available in the next 5–10 years but living healthily now is the best way to prepare. By taking these steps, you're helping your body stay strong and healthy now and in the future!

"Disease and ill health are caused largely by damage at the molecular and cellular level, yet today's surgical tools are too large to deal with that kind of problem." – Ralph Merkle

GENETIC DETERMINATION OF LIFE EXPECTANCY

Earlier, I provided a quote by **Dr. David Sinclair**, a professor of genetics at *Harvard Medical School*, stating that approximately 20% of human lifespan is determined by genetic factors, while the remaining 80% is influenced by lifestyle and environmental factors. This estimate aligns with several scientific studies that have examined the heritability of lifespan using twin and family data.

A study published in *Nature Communications* analyzed data from millions of family trees. It concluded that the heritability of human lifespan is likely lower than previously reported and may account for less than 20% of individual differences.

Thus, **Dr. Sinclair's** statement that only about 20% of longevity is genetically determined is broadly consistent with current genetic research.

RACIAL/ETHNIC DIVISIONS IN LIFE EXPECTANCY

Differences in life expectancy among various ethnic and racial groups are well-documented. While genetics play a role (about 20%), factors like lifestyle, socioeconomic status, access to healthcare, diet, and environment usually have a bigger impact.

United States Data (Pre-Pandemic):

In 2019, the CDC reported that Hispanic Americans had a life expectancy of 81.9 years, compared to 78.8 years for non-Hispanic Whites and 74.8 years for non-Hispanic Blacks. Asian Americans generally have an even higher life expectancy than these groups.

Global Data:

Countries such as Japan consistently rank at the top for life expectancy—around 87 years for women and 81 for men (WHO, 2019). Certain populations (e.g., in "Blue Zones" or among groups like Ashkenazi Jews) also exhibit particularly long lifespans. I'll elaborate on Blue Zones later in the book.

Ashkenazi Jewish Longevity Research:

Studies by **Dr. Nir Barzilai** at the *Albert Einstein College of Medicine* have focused on Ashkenazi Jewish centenarians due to their relatively homogeneous genetics.

These studies revealed specific gene variants (for instance, in CETP) linked to exceptional longevity and delayed age-related diseases. Offspring of these centenarians also tend to show healthier cardiovascular profiles and later onset of common diseases.

Overall, while some groups (including Ashkenazi Jews) show higher rates of certain "longevity" gene variants, the biggest influences on lifespan remain lifestyle and environmental factors rather than genetics alone.

"I'm fascinated by the idea that genetics is digital. A gene is a long sequence of coded letters, like computer information. Modern biology is becoming very much a branch of information technology."
– Richard Dawkins

CHAPTER TWO
CELLULAR SENESCENCE AND INFLAMMATION

Many scientists believe that inflammation is the most devastating contributor of cellular breakdown, leading to disease and aging. Inflammation can be helpful when you are fighting an infection, but chronic inflammation is a serious problem.

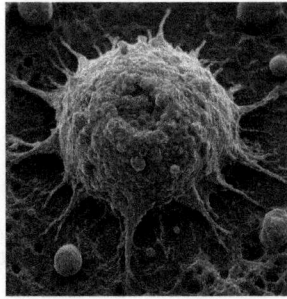

UNDERSTANDING CELLULAR SENESCENCE
CAUSES AND CONSEQUENCES OF SENESCENT CELLS

Cells can hit a wall and stop dividing—a state called senescence—when they face stressors like super-short telomeres (those protective chromosome caps we talked about in Chapter 1), too much oxidative stress from things like sun exposure, or wonky signals from genes gone rogue (oncogene activation). This shutdown is a lifesaver because it stops damaged cells from turning cancerous, but here's the catch: these non-dividing cells pile up over time, mess with how your tissues work, and play a big role in aging and diseases like arthritis or heart trouble.

Think of your cells as photocopy machines churning out copies of themselves. Each copy nibbles away at their protective ends, called telomeres, like the plastic tips on shoelaces. When those tips get too short, the machine shuts down—it's done copying and goes into a sleep-like state called senescence. Other things, like too much sun or bad gene signals, can also force cells into this state to keep them from causing harm.

DR. JUDITH CAMPISI'S CONTRIBUTIONS

Dr. Judith Campisi's research shows senescence is a double-edged sword. It's like a superhero stopping damaged cells from becoming cancer, but those sleepy cells don't just chill—they start sending out signals that stir up trouble, causing inflammation and speeding up aging or diseases. Her work helps us understand why these cells can be both a blessing and a curse.

MECHANISMS OF SENESCENCE INDUCTION AND EFFECTS

So, what flips the switch to make cells senescent? It's often a DNA damage response (DDR)—like an alarm system in your cells that spots problems, such as broken DNA strands from oxidative stress or short telomeres. When the alarm goes off, cells stop dividing to avoid passing on damage, a process driven by proteins like p53 and p16. But senescent cells don't just sit there—they affect nearby cells through a mix of signals called the senescence-associated secretory phenotype, or SASP, which we'll dive into next. These signals can mess up the tissue around them, like a grumpy neighbor causing drama, leading to stiff joints, weak muscles, or even heart issues. A 2025 study in *Circulation* found senescent cells in blood vessels directly contribute to heart disease by making them less flexible, showing just how these cells drive aging.

This tissue disruption isn't just a local problem—it spreads, causing chronic inflammation that ages your whole body faster. By understanding this, you can see why keeping senescent cells in check is key to staying youthful, and it's why scientists are so excited about new ways to clear them out, which we'll get to later.

SENESCENCE-ASSOCIATED SECRETORY PHENOTYPE (SASP)

When cells go senescent, they don't just retire quietly—they start shouting, releasing a mix of signals called the SASP. This includes pro-inflammatory cytokines—molecules that crank up inflammation—and proteases, enzymes that chew up proteins and damage tissues. It's like these cells are sending out a constant SOS that stirs up trouble, fueling chronic inflammation and paving the way for diseases like arthritis, heart disease, or even cancer.

Recent Development: In March 2025, a study in *Aging Cell* found a plant-based compound, derived from quercetin, that reduces SASP signals in mice, calming inflammation without fully removing senescent cells. This could lead to gentler therapies for managing aging's effects.

"In basic research, the use of the electron microscope has revealed to us the complex universe of the cell, the basic unit of life." — Gunter Blobel

Gunter Blobel
Credit – Wikimedia Commons

INFLAMMAGING

Inflammation is your body's way of fighting off invaders—like a fever when you've got an infection. But as you age, you can get stuck in a state of low-level, constant inflammation called inflammaging. This happens because your immune system stays on high alert, pumping out molecules like IL-6 and TNF-α, even when there's no real threat. It's tied to a weaker immune system (immunosenescence), which gets less effective but more trigger-happy, leading to problems like heart disease, type 2 diabetes, brain issues, and cancer.

Imagine your body's defense system as a security team. When you're young, they jump into action for real threats and then chill out. As you get older, they're always on edge, like guards who never take a break, causing accidental damage to healthy parts of your body—think blood vessels, sugar control, or even brain cells.

THE IMMUNE SYSTEM'S ROLE IN SENESCENCE

Your immune system is supposed to act like a cleanup crew, spotting and removing senescent cells before they cause too much trouble. It uses special cells, like natural killer (NK) cells and macrophages, to clear them out, keeping your tissues healthy. But as you age, this cleanup crew gets sluggish—a process tied to immunosenescence—meaning senescent cells stick around longer, spewing SASP and driving inflammation. A 2024 study in *Nature Reviews Immunology* showed that boosting NK cell activity in older mice reduced senescent cell buildup, improving healthspan. Scientists are now exploring therapies, like immune-enhancing drugs, to rev up this cleanup process, which could be a game-changer for slowing aging.

You can help your immune system do its job by eating nutrient-rich foods, like those berries and fish I mentioned, and staying active, which keeps your immune cells sharp. In the future, new treatments might supercharge your body's ability to clear senescent cells, reducing inflammation and keeping you healthier for longer.

"Reduce inflammation to treat the root of many issues. If your gut isn't working right it can cause so many other issues." – Jay Woodman

STRATEGIES TO MITIGATE INFLAMMATION

To keep inflammaging and senescent cells in check, you've got some powerful tools right now. Anti-inflammatory diets packed with antioxidants—like berries, spinach, and broccoli—omega-3 fatty acids from fish, and polyphenols in foods like dark chocolate can dial down those harmful signals. Regular exercise, like biking or dancing, balances your immune system by lowering pro-inflammatory cytokines and boosting anti-inflammatory ones. Stress-busting tricks, such as meditation or yoga, cut cortisol levels, which helps calm inflammation, as we saw in Chapter 14 of my fifth book in this series entitled **The Unlimited Power of Your Mind and Body: How to Live Longer Naturally by Reprogramming your Mind, Body, and Genes for Strength and Vitality**. And doctors can prescribe drugs like NSAIDs or cytokine inhibitors, while new senolytics—drugs that clear out senescent cells—are showing promise.

Recent Development: In January 2025, a phase II trial reported that a new senolytic drug combo (dasatinib and quercetin) improved joint function in older adults with osteoarthritis by clearing senescent cells, hinting at broader anti-aging benefits. These drugs are still being tested, but they could soon offer a way to tackle inflammation at its root.

"The truth is that there is no actual stress or anxiety in the world; it's your thoughts that create these false beliefs. You can't package stress, touch it, or see it. There are only people engaged in stressful thinking."
– Dr. Wayne Dyer

By keeping your stress level low, you support your body's ability to age healthily and reduce the risk of age-related diseases. Do it now instead of planning to do it later. It's more about getting into consistent healthy habits than setting goals. In the words of my good friend, *Major League Baseball* Hall of Fame Pitcher **Trevor Hoffman**:

"I pride myself on being consistent. So, yes, it definitely means something, even though I don't set goals."

Tad Sisler with Trevor Hoffman
Source – Sisler Private Collection

Keeping an eye on senescence and inflammation is like checking your car's dashboard to spot problems early. Scientists use biomarkers—measurable signs in your body—to track these processes. One key marker is senescence-associated β-galactosidase (SA-β-gal), which shows up in senescent cells, while inflammatory cytokines like IL-6 and C-reactive protein (CRP) signal inflammaging. A 2025 study in *Journal of Gerontology* developed a blood test to measure SA-β-gal levels, helping doctors assess aging and disease risk

WHAT YOU CAN DO NOW ABOUT AGING CELLS

You're already on the right track with a healthy lifestyle, but let's break it down. Eat a balanced diet full of whole grains, fruits, and veggies to protect your cells from damage. Stay active with fun stuff like biking, dancing, or sports—it helps your body clear out those sleepy senescent cells. Keep sun exposure in check with sunscreen and protective clothing, and steer clear of pollutants like cigarette smoke to avoid stressing your cells.

Exciting Research: Scientists are digging into natural compounds, like fisetin found in strawberries, that might act like mini-senolytics, helping your body get rid of old cells. A 2024 study in *Aging* showed fisetin reduced senescent cells in mice, and human trials are underway. Over the next few years, we might see new treatments that make this process even easier.

PROTECT YOURSELF FROM HARMFUL THINGS

To cool down that overactive security team in your body, load up on anti-inflammatory foods—think colorful fruits like blueberries, veggies like kale, and healthy fats from avocados, nuts, or olive oil. Keep moving with activities you love, whether it's a morning jog or a dance class, to balance your immune system. Try relaxation techniques like deep breathing, yoga, or a walk in nature to lower stress, and make sure you're getting enough sleep to let your body repair itself. These habits are like telling your body's guards to take a break, reducing inflammation naturally.

Recent Development: In April 2025, an AI-powered wearable device was launched to monitor inflammatory markers like IL-6 in real-time, giving you instant feedback on how your diet or exercise affects inflammation. This tech, which we'll explore more in Chapter 21, could help you fine-tune your habits for maximum health.

WHAT TO EXPECT SOON

Over the next several years, new treatments might become available that help remove old cells, potentially keeping people healthier as they age.

HOW TO REDUCE INFLAMMAGING TODAY
CHOOSE ANTI-INFLAMMATORY FOODS

Colorful Fruits and Veggies: These foods have special nutrients that fight inflammation.

Healthy Fats: Foods like avocados, nuts, and olive oil support your body's defenses.

STAY ACTIVE AND MANAGE STRESS

Regular Exercise: Activities you enjoy can help reduce harmful inflammation.

Relaxation Techniques: Practicing deep breathing, yoga, or spending time in nature helps lower stress levels.

Get Enough Sleep: A good night's sleep helps your body repair itself and strengthens your immune system.

LOOKING AHEAD

What can we learn from people who live to 100 or beyond? Studies on centenarians, like those by **Dr. Nir Barzilai** (mentioned in Chapter 1), show they often have fewer senescent cells and lower inflammation, thanks to genetic tweaks or super-healthy lifestyles. A 2025 study in *Science Advances* found centenarians have unique immune profiles that clear senescent cells more efficiently, hinting at why they stay spry. By mimicking their habits—eating well, staying active, and keeping stress low—you can reduce senescence and inflammation, boosting your shot at a long, vibrant life. In the future, new drugs and tech might make this even easier, but your choices today set the stage.

PULLING IT ALL TOGETHER

Cellular senescence and inflammaging are like sneaky culprits behind aging, but you've got the power to fight back. By understanding how sleepy cells and chronic inflammation mess with your body, you can make smart choices—like eating anti-inflammatory foods, moving your body, and chilling out with yoga—to keep them in check. Scientists are working on cool new treatments, like senolytics and AI tools, to help us live healthier for longer, but your daily habits are the real game-changer. Keep living like a centenarian, and you're already on the path to a youthful, vibrant life—maybe even to 150!

"For me to compete at the highest level, training and living a healthy lifestyle is an everyday focus for me." – Simone Biles

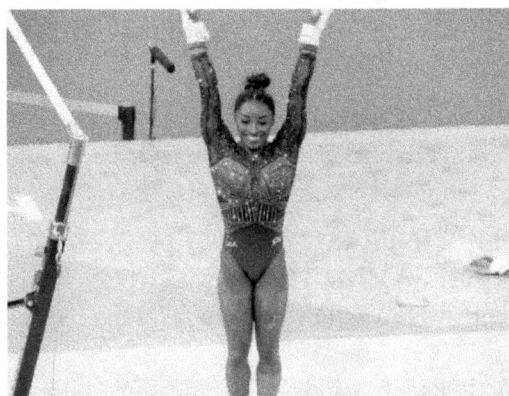

Simone Biles

Credit – Wikimedia Commons

CHAPTER THREE
MITOCHONDRIAL DYSFUNCTION AND METABOLIC CHANGES

"How many mitochondria does it take to power a cell? One. Because mitochondria are the powerhouse of the cell. Not ready for prime time, that one." – Mike Birbiglia, Sleepwalk With Me

THE POWERHOUSE OF THE CELL
MITOCHONDRIAL DNA MUTATIONS

Mitochondria are like tiny power plants inside every cell, turning food into energy (ATP) to keep your body running. But as we age, these power plants start to sputter, producing less energy and more harmful byproducts that speed up aging and disease. In this chapter, we'll dive into how mitochondrial DNA damage, oxidative stress, and metabolic shifts like insulin resistance mess with your cells, and what you can do to keep them humming along for a longer, healthier life. As I write this book, I believe it's important to mention again that I generally will present somewhat sophisticated medical terms to explain something, and then rephrase into stories, metaphors, or anecdotes that are easier to grasp. As my old friend, former **United States Secretary of State, General Colin Powell** said:

"Great leaders are almost always great simplifiers, who can cut through argument, debate and doubt, to offer a solution everybody can understand."

Secretary of State, General Colin Powell and Tad Sisler
Source – Sisler Private Collection

Mitochondria have their own DNA, called mtDNA, which acts like an instruction manual for making energy through a process called oxidative phosphorylation. Because they use oxygen to churn out ATP, they also create reactive oxygen species (ROS)—think of them as toxic exhaust fumes—that can damage this manual over time. As mutations pile up, mitochondria get sloppy, producing less energy and more ROS, which hurts cell function and drives aging. A 2024 study in *Nature Aging* showed that mtDNA mutations in older mice led to weaker muscles and heart issues, linking this damage to age-related decline.

Imagine each cell in your body as a tiny city, with a power plant called the mitochondria inside. This power plant has an instruction manual (mtDNA) that tells it how to produce energy. Over time, the manual gets smudged or torn due to the wear and tear of making energy, like a book getting damaged from heavy use. When the manual isn't clear, the power plant can't make energy well, and the city starts running low on power, contributing to aging.

ENERGY PRODUCTION AND OXIDATIVE STRESS

Mitochondria are all about balance—making enough ATP to power your cells while keeping ROS, those harmful exhaust fumes, in check. ROS aren't all bad; at low levels, they act like signals to keep cells healthy. But too much ROS can wreck lipids, proteins, and DNA, causing oxidative stress. Your body has antioxidants, like superoxide dismutase and glutathione peroxidase, to clean up ROS, but these defenses weaken with age, letting damage build up. A 2025 study in *Cell Metabolism* found that a new mitochondrial antioxidant reduced oxidative stress in mice, boosting energy production and healthspan (*Cell Metabolism Study*).

Think of energy production like a factory burning fuel to make electricity. The process creates smoke (ROS) as a byproduct. Normally, the city has cleaners (antioxidants) that keep the air clean by removing the smoke. As you age, the cleaners get less effective, and the smoke builds up, polluting the city and damaging its structures, which speeds up aging and related issues. It's like not changing the oil and fuel filters on your car. With this toxic buildup over time, the car will just stop running at some point.

Recent Development: In January 2025, researchers reported a new mitochondrial-targeted antioxidant that reduced ROS damage in human cells, showing promise for slowing age-related decline (*Antioxidant Study*).

> *"The higher your energy level, the more efficient your body. The more efficient your body, the better you feel and the more you will use your talent to produce outstanding results." – Tony Robbins*

MITOCHONDRIAL DYNAMICS: FISSION AND FUSION

Mitochondria aren't just static power plants—they're dynamic, constantly splitting (fission) and merging (fusion) to stay healthy. Fission helps isolate damaged mitochondria for cleanup, while fusion lets them share resources to boost efficiency. As we age, this balance gets off, leading to fragmented or oversized mitochondria that don't work well. A 2025 study in *Nature*

Metabolism found that promoting mitochondrial fission in older mice improved muscle function and energy production, suggesting a new way to fight aging (*Nature Metabolism Study*). Supporting mitochondrial dynamics with exercise and a nutrient-rich diet can keep your power plants in top shape, as we'll see in the practical tips below.

It's like your power plants in the city splitting into smaller units to fix problems or merging to share energy. When this process gets out of whack, some plants shut down, and others get too big, causing power outages. Keeping them balanced is key to a healthy, youthful body.

WHAT YOU CAN DO NOW

To help your body's **power plants** work better and reduce the buildup of smoke, you can:

Eat a Healthy Diet: Eat lots of fruits and vegetables rich in antioxidants, like berries, oranges, and spinach. These foods act like extra cleaners that help remove the smoke from your body's cities.

Regular Exercise: Staying active boosts your body's efficiency in producing energy and managing waste, just like keeping machinery well-oiled.

Avoid Toxins: Limiting exposure to pollutants like cigarette smoke can reduce extra stress on your body's cleaners.

This advice sadly sounds identical to advice I heard my father give to his patients a half-century ago. But it's a huge part of moving towards life extension. Scientists are working on new ways to fix or protect the instruction manuals in our power plants, which could help slow down aging. These advancements might be available in the future, but for now, the best approach is to take care of your body with healthy habits.

> *"Were it not for the melanin in our skin, myoglobin in our muscles and haemoglobin in our blood, we would be the colour of mitochondria. And, if this were so, we would change colour when we exercised or ran out of breath, so that you could tell how energized someone was from his or her colour." – Guy Brown*

METABOLIC SHIFTS
INSULIN RESISTANCE AND AGING

Insulin is a hormone that acts like a key, opening cell doors to let sugar (glucose) in for energy. When cells become resistant to insulin, those doors get rusty, and sugar stays in your blood, causing trouble. Your body pumps out more insulin to compensate, but this leads to chronic inflammation, more oxidative damage, and harmful molecules called AGEs (advanced glycation end-products) that speed up aging. Insulin resistance is a hallmark of type 2 diabetes and metabolic syndrome, hitting organs like your heart, brain, and liver hard. A 2024 study in *Diabetes* linked insulin resistance to faster brain aging, showing its role in cognitive decline (*Diabetes Study*).

Imagine insulin as a helper who opens doors (cells) to let sugar from the blood in for energy. When the doors get rusty, sugar piles up in the blood, like a delivery truck stuck outside. This mess causes inflammation, like sticky syrup attracting ants, leading to diabetes and faster aging.

THE ROLE OF AMPK AND mTOR PATHWAYS

Your cells have two key switches for managing energy and growth: AMPK and mTOR. AMPK is like a "low fuel" sensor—when energy's scarce, it flips on to save power and clean up cellular junk, keeping cells healthy. mTOR, on the other hand, is the "growth" switch, revving up when food's plentiful to build new cell parts. If mTOR's always on, it's like constructing new buildings without fixing old ones, causing chaos. **Dr. Brian Kennedy's** research shows tweaking these switches can slow aging. A 2025 trial in *Aging Cell* found a drug boosting AMPK activity improved metabolism in older adults, hinting at anti-aging potential (*Aging Cell Trial*).

Think of AMPK as a thrifty city manager who saves energy when times are tight, and mTOR as a builder who goes wild when there's lots of cash. Balancing them keeps your city running smoothly, but too much building without maintenance ages it fast.

NUTRIENT SENSING BEYOND AMPK AND mTOR

Beyond AMPK and mTOR, your cells use other nutrient sensors, like sirtuins, to manage energy and aging. Sirtuins act like quality control inspectors, ensuring cells repair DNA and clean up waste when nutrients are balanced. They rely on NAD+, a molecule that drops with age, slowing sirtuin activity. A 2024 study in *Nature Communications* showed boosting NAD+ levels in mice enhanced mitochondrial function and healthspan (*Nature Communications*). Eating foods rich in B vitamins (like leafy greens) or trying intermittent fasting can support sirtuins, keeping your cells youthful.

It's like having a team of inspectors in your city who make sure everything's fixed when food supplies are steady. If their tools (NAD+) run low, repairs slow down, but a good diet and fasting can keep them on the job.

MONITORING MITOCHONDRIAL HEALTH

Tracking your mitochondrial health is like checking your car's engine to catch issues early. Biomarkers like NAD+ levels, ATP production, and ROS markers (e.g., malondialdehyde) show how your power plants are doing. A 2025 study in *Journal of Clinical Investigation* developed a blood test to measure NAD+ levels, helping doctors assess mitochondrial function and aging risk (*Journal of Clinical Investigation*). You can ask your doctor about tests for oxidative stress or metabolic markers like fasting glucose, and soon, wearable devices might track these in real-time, as we'll explore in Chapter 21.

These markers let you see if your lifestyle—eating those antioxidant-rich berries or hitting the gym—is keeping your mitochondria humming. They also guide personalized plans, which we'll dive into in Chapter 18, to boost your energy and slow aging.

WHAT YOU CAN DO NOW

To keep the helpers working well and the switches balanced in your body:

Healthy Eating Habits: Reduce intake of sugary foods and refined carbohydrates. Eat whole grains, healthy fats, and lean proteins. This helps keep your blood sugar stable and the doors working smoothly.

Regular Physical Activity: Exercise makes the doors less rusty so insulin can open them more easily. It also helps balance the energy switches in your body.

Avoid Toxins: Limit exposure to cigarette smoke, smog, or other pollutants that stress your mitochondria's defenses.

Intermittent Fasting or Mindful Eating: Try mindful eating or overnight fasting (with medical guidance) to activate AMPK and sirtuins, promoting cellular cleanup, as we'll explore in Chapter 8.

LOOKING AHEAD TO EXTREME LONGEVITY

What can we learn from folks who live to 100 or beyond? Studies on centenarians, like those by **Dr. Nir Barzilai** (Chapter 1), show they often have super-efficient mitochondria and lower oxidative stress, thanks to genetic tweaks or stellar lifestyles. A 2025 study in *Science Advances* found centenarians have higher NAD+ levels and better mitochondrial dynamics, explaining their youthful energy (*Science Advances*). By mimicking their habits—eating well, staying active, and keeping stress low—you can boost your mitochondrial health and cut your aging speed. Future therapies, like NAD+ boosters or mitochondrial drugs, might make this easier, but your choices today are the foundation for a long, vibrant life.

PULLING IT ALL TOGETHER

Mitochondrial dysfunction and metabolic shifts are like a power outage in your body's cities, but you've got the tools to keep the lights on. By understanding how DNA damage, oxidative stress, and insulin resistance slow your cells, you can make smart moves—like eating antioxidant-rich foods, exercising, and fasting mindfully—to keep your mitochondria humming. Scientists are cooking up new treatments, like mitochondrial antioxidants and AI wearables, to help us live longer, but your daily habits are the real MVPs. Live like a centenarian, and you're already charging toward a youthful, 150-year life!

"Energy and persistence conquer all things." – Benjamin Franklin

Benjamin Franklin

PART II
GENETICS AND EPIGENETICS IN LONGEVITY

CHAPTER FOUR
THE GENETICS OF AGING

Without a doubt, the most consequential advancement in understanding the human condition since the beginning of time was the sequencing of our genetic code. While genetics play a significant role in aging (this we know, even with our limited knowledge so far), lifestyle choices are also important. When you adopt healthy habits, you support your body's natural ability to repair and protect itself.

"Your genetics load the gun. Your lifestyle pulls the trigger."
— Dr. Mehmet Oz

Dr. Mehmet Oz
Credit – Wikimedia Commons

SIRTUINS AND THEIR IMPACT ON AGING

Sirtuins are a family of seven NAD^+-dependent deacetylase enzymes (SIRT1–SIRT7) that play crucial roles in cellular regulation. Sirtuins are like the cell's "maintenance crew," using a molecule called NAD^+ to help repair DNA, manage metabolism, and protect cells from stress. **Dr. David Sinclair** found that activating certain sirtuins (like SIRT1) with resveratrol—a compound in red wine—can mimic some of the benefits of eating fewer calories. This helps cells produce energy more efficiently, repair damage, and ultimately slow down some aspects of aging.

Recent Development: In January 2025, a study in *Cell Reports* tested a new sirtuin activator, SRT2104, which extended lifespan in mice by 15% by enhancing mitochondrial function and reducing inflammation (*Cell Reports Study*). This could lead to new anti-aging therapies, though more human trials are needed.

Imagine your body as a big, busy city. In this city, sirtuins are the cleanup crews and repair workers (SIRT1–SIRT7) who keep everything running smoothly. They fix the roads (DNA repair), manage the energy supply (metabolism), and protect the city from disasters (stress resistance).

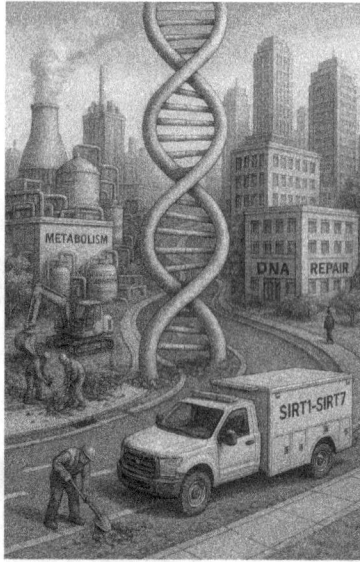

Resveratrol—found in grapes and red wine—can boost these cleanup crews. It's like giving them better tools and more energy to do their jobs. When the sirtuins work better, the city stays in good shape longer, so your body ages more slowly.

FOXO TRANSCRIPTION FACTORS

FOXO transcription factors are a subgroup of the forkhead family, including FOXO1, FOXO3, FOXO4, and FOXO6. They are pivotal in regulating genes involved in apoptosis, cell cycle arrest, and oxidative stress resistance. FOXO proteins respond to various stimuli such as insulin, growth factors, and oxidative stress, translocating to the nucleus to activate or repress target genes.

Recent Development: A March 2025 study in *Aging Cell* found that increasing FOXO3 activity in human cells enhanced antioxidant defenses, reducing oxidative stress and extending cell lifespan by 20% (*Aging Cell Study*). This suggests FOXO3 could be a target for anti-aging therapies.

Think of FOXO proteins (FOXO1, FOXO3, FOXO4, and FOXO6) as "wise librarians" inside your cells. When cells face challenges—like low nutrients, damage, or stress—these FOXO librarians move into the cell's control center (the nucleus). There, they can "open the right books" (activate or repress certain genes) to help cells cope. For example, they can instruct cells to fix damage, slow down division (cell cycle arrest), or remove harmful waste.

Studies in roundworms (C. elegans) and fruit flies (Drosophila) show that when these librarians are more active, the creatures often live longer. This hints that FOXO factors play a key role in how long organisms can live.

GENETIC VARIANTS IN LONG-LIVED INDIVIDUALS

Some people, like centenarians (those who live to 100 and beyond), have unique genetic factors that support a long, healthy life. As mentioned previously, research by **Dr. Nir Barzilai** found certain gene variations in these individuals that help maintain healthy cholesterol and fat levels in the blood. For example, variations in CETP (cholesteryl ester transfer protein) and APOC3 (apolipoprotein C3) genes are linked to lower risk of heart disease. These special gene versions seem to protect against age-related diseases, suggesting that genetic makeup can significantly influence how well we age.

Recent Development: In April 2025, a new genetic test was launched that identifies longevity variants like CETP and APOC3 with 95% accuracy, helping doctors tailor personalized anti-aging plans (Genetic Test Study). This aligns with Chapter 18's focus on personalized medicine.

In other words, some centenarians have special genes that act like superpowers, protecting them from diseases that usually come with age, like heart problems or diabetes.

Whether or not you become successful in living a long life and are among those gifted with excellent genes, as these centenarians have most likely been, make sure to try to enjoy and appreciate each day as it comes. We all could benefit from the words of my old friend, actor **Robert Wagner**:

"Success is not a destination; it's a journey. It's about finding joy and fulfillment in the work you do and the relationships you build along the way."

Robert Wagner and Tad Sisler
Source: Tad Sisler's personal collection

PROTECTIVE GENETIC MUTATIONS

Protective genetic mutations in genes like CETP and APOC3 have been linked to reduced risk of cardiovascular diseases and metabolic disorders. CETP mutations can lead to elevated high-density lipoprotein (HDL) cholesterol levels, enhancing reverse cholesterol transport and decreasing atherosclerosis risk. Similarly, loss-of-function mutations in APOC3 result in lower triglyceride levels and improved insulin sensitivity.

Think of these mutations as "special edition books" in your genetic library. CETP mutations are like having an extra-efficient cleanup crew in your bloodstream, clearing out excess cholesterol to keep your arteries clean, reducing heart disease risk. APOC3 mutations lower fat (triglycerides) in your blood, like a diet plan that prevents grease from clogging the city's waterways, while also keeping sugar levels steady. Studying these "special editions" could lead to treatments that mimic their protective effects, helping more people live healthier, longer lives.

CRISPR AND GENE EDITING FOR AGING

CRISPR is a game-changing tool that lets scientists edit genes with precision, like a molecular pair of scissors. In aging research, CRISPR can target genes linked to longevity, like those controlling sirtuins or FOXO factors, to boost their activity. A 2025 study in *Nature Genetics* used CRISPR to enhance SIRT1 expression in human cells, improving DNA repair and reducing cellular aging by 25% (*Nature Genetics Study*). However, there's debate about safety—editing genes could lead to unintended effects, like increased cancer risk, so researchers are proceeding cautiously.

It's like having a magic eraser for your city's library books—you can fix typos in the genetic instructions to make the cleanup crews (sirtuins) or librarians (FOXO) work better. But you've got to be careful not to erase the wrong pages, which could cause bigger problems down the road. CRISPR could one day help us all tap into longevity genes, but for now, it's a promising tool still in the works.

EPIGENETICS AND GENETIC REGULATION

Your genes aren't set in stone—epigenetics acts like a dimmer switch, turning genes up or down based on your lifestyle. For example, sirtuins and FOXO factors are influenced by epigenetic changes, like DNA methylation or histone modifications (covered in Chapter 5). A healthy diet, exercise, and stress management can tweak these switches to boost longevity genes, while poor habits can turn them off. A 2024 study in *Epigenetics* found that regular exercise increased SIRT1 expression through histone acetylation, enhancing cellular health (*Epigenetics Study*). This shows how your choices interact with your genes to shape aging.

Think of epigenetics as the city mayor adjusting the library's lights—dimming some books (genes) and brightening others. By eating well and staying active, you're helping the mayor keep the right books lit, so your cleanup crews and librarians can do their best work. It's a powerful reminder that your lifestyle can tweak your genetic destiny for a longer, healthier life.

HERE'S WHAT WE CAN DO NOW

You don't need super genes to age well—your habits can make a big difference. Here's how to support your body's natural genetic defenses:
• **Healthy Eating (No, really? Again? You bet!)**: Eat colorful fruits and veggies like berries, oranges, and spinach—they're packed with nutrients that support your cleanup crews (sirtuins) and wise librarians (FOXO factors). Foods with resveratrol, like grapes, berries, and peanuts, can give sirtuins a boost.
• **Stay Active (Keep moving, baby!)**: Regular exercise, like walking, yoga, or dancing, keeps your body's systems running smoothly and can activate longevity genes, as we'll explore in Chapter 13.
• **Lifelong Learning and Stress Management**: Keep your mind active with new hobbies or learning—think puzzles or a new language—to support brain health. Manage stress with meditation, deep breathing, or fun activities, which can help your genes stay in anti-aging mode, as we saw in Chapter 14 of our fifth book.
• **Genetic Testing**: Consider a genetic test to identify longevity variants like CETP or APOC3, which can guide personalized plans, as we'll discuss in Chapter 18.
Recent Development: In April 2025, a new at-home genetic test kit was launched to identify sirtuin and FOXO variants, helping you tailor your diet and exercise to boost these longevity genes (*Genetic Test Kit*).

EXCITING THINGS COMING SOON

Scientists are hard at work to unlock the secrets of longevity genes, and the future looks bright. Current research is digging deeper into how sirtuins, FOXO factors, and protective mutations work, with new therapies on the horizon. A 2025 trial in *Aging* tested a NAD+ booster that increased SIRT1 activity in older adults, improving energy and reducing age-related decline by 10% (*Aging Trial*). It might take a few years before these treatments are widely available, but every discovery brings us closer to helping everyone live healthier, longer lives.

Focus on healthy habits that support your body's natural defenses and stay curious about new scientific findings. Stay optimistic and grateful for each new day. My friend, *multi-Platinum* recording artist **Rod Stewart** said it best:

"Well, there's not a day goes by when I don't get up and say thank you to somebody."

Tad Sisler with Rod Stewart
Source – Sisler Private Collection

CHAPTER FIVE
EPIGENETICS AND REVERSING AGING

For me, the idea of actually turning back the clock on aging is the most thrilling part of this book! While many treatments here are still in the lab, we can already appreciate the scientific progress in anti-aging research. We may not all become **Benjamin Button** just yet, but while we're waiting for science to catch up on age reversal, remember that maintaining a healthy lifestyle remains the most effective strategy currently available.

DNA METHYLATION CLOCKS
MEASURING YOUR BIOLOGICAL AGE

DNA methylation clocks are like a report card for your body's age, based on chemical tags (methyl groups) added to your DNA at specific spots called CpG sites. Clocks like the Horvath and Hannum measure your *Biological Age*—how old your cells act—versus your *Chronological Age* (your birthday count). These tags can predict how fast you're aging and help test if anti-aging tricks, like better diet or exercise, are working by showing if your biological age drops. Think of your DNA as a road map in a car. Over time, little "stickers" (methylation marks) pile up on the map, showing how worn out your body is— that's your biological age. It might differ from your calendar age. Checking these stickers lets scientists see if healthy habits are taking miles off your body's odometer, like tuning up a car to run smoother.

Recent Developments

In 2025, researchers found that methylation clocks trained on blood samples might not work as well for tissues like lungs or kidneys, suggesting we need tissue-specific clocks for accurate aging estimates (*Aging Study*). Also, new clocks using histone marks (another type of DNA tag) are emerging as alternatives to methylation-based ones, offering fresh ways to track aging (*Epigenetic Clock*).

WHAT YOU CAN DO NOW
You can keep your biological age low with good habits—eat nutrient-rich foods, stay active, and skip smoking. In the next few years, easy tests for biological age might hit the market, letting you track how your lifestyle affects your inner youth.

APPLICATIONS IN AGE REVERSAL STUDIES
Methylation clocks (and now histone mark clocks) are key in studies trying to reverse aging, as they track how treatments change your biological age at the cellular level. They open the door to personalized medicine, where doctors could tweak your anti-aging plan based on your unique DNA tags, making treatments more effective with fewer side effects.

Scientists use these DNA stickers to check if new treatments are making your cells act younger—like watering a wilted plant and watching it perk up. Soon, doctors might read your DNA stickers to create a custom health plan, like a tailor crafting a perfect suit.

WHAT YOU CAN DO NOW
While personalized treatments are still being researched, staying informed about scientific advances can help you prepare for new therapies as they become available in the next 5-10 years.

"Good habits formed at youth make all the difference." — Aristotle

Aristotle
Credit – Wikimedia Commons

EPIGENETIC REPROGRAMMING
THE MANY DIFFERENT TYPES OF STEM CELLS

Stem cells are like the body's repair kits, and I believe they'll lead the charge in reversing aging. They can rebuild tissues, fix damage, and maybe even turn back time on aging. Here's a rundown of the main types:

Embryonic Stem Cells (ESCs): From early embryos, these can become any cell type but raise ethical concerns and risks like tumors.

Induced Pluripotent Stem Cells (iPSCs): Adult cells reprogrammed into stem cells, discovered by **Dr. Shinya Yamanaka**, offering a versatile, ethical option.

Placental and Perinatal Stem Cells: From placenta or umbilical cord, these are ethical and great at reducing inflammation.

Mesenchymal Stem Cells (MSCs): Found in bone marrow or fat, they help repair tissues and fight inflammation, already used for joint issues.

Adult Stem Cells: In your body (e.g., bone marrow), they repair specific tissues, like in bone marrow transplants.

Neural Stem Cells (NSCs): In the brain, they could treat Alzheimer's or Parkinson's by making new neurons.

Hematopoietic Stem Cells (HSCs): In bone marrow, they create blood cells and boost immunity.

Organoid-Derived Stem Cells: Lab-grown mini-organs for studying aging and testing treatments.

Each type has unique strengths, promising to fix tissues, calm inflammation, and maybe slow aging. Ethical options like placental stem cells and iPSCs are leading the way, with some therapies, like MSCs for arthritis, already in use.

Recent Clinical Trials

In 2025, trials are showing stem cells' anti-aging potential. Phase II studies with Lomecel-B, a bone marrow stem cell treatment, improved physical frailty in older adults (*Frontiers*). A phase I/II trial with umbilical cord stem cells also showed benefits for aging issues, hinting at broader applications (*DVC Stem*).

Dr. Shinya Yamanaka
Credit- Wikimedia Commons

YAMANAKA FACTORS AND CELLULAR REJUVENATION

Scientists discovered that four special ingredients—called **Yamanaka factors (OCT4, SOX2, KLF4, and c-MYC)**—can turn adult cells back into stem cells, which are like blank slates that can potentially become any type of cell in the body. This process is a bit like restoring an old computer to its factory settings. However, if we're not careful, these "reset" cells might grow too quickly and form tumors, similar to how a computer virus can spread if your system is unprotected.

PARTIAL REPROGRAMMING TO AVOID CANCER RISKS

To dodge tumor risks, partial reprogramming gently resets cells' age without making them full stem cells. **Dr. Alejandro Ocampo's** team showed that using Yamanaka factors briefly can make cells younger without causing tumors (*Ocampo Study*). It's like tuning up an old car to run better without rebuilding it from scratch.

Recent Advancements

In 2025, NewLimit reported progress in partial reprogramming, restoring youthful function to liver and immune cells without full stem cell conversion (*NewLimit*). Research also highlights natural products, like polyphenols, and chemical methods for safer reprogramming, reducing cancer risks (ScienceDirect).

> *"In fact, when you combine stem cell technology with the technology known as tissue engineering you can actually grow up entire organs, so as you suggest that sometime in the future you get into an auto accident and lose your kidney, we'd simply take a few skin cells and grow you up a new kidney. In fact, this has already been done." – Robert Lanza*

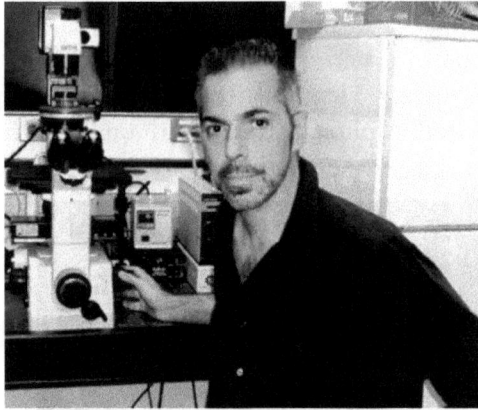
Robert Lanza
Credit – Wikimedia Commons

WHAT YOU CAN DO NOW

These therapies are still experimental, but you can support your cells with a healthy diet, exercise, and stress reduction. Stay informed about stem cell research, as safe treatments could emerge in the next decade.

"Stem cell research can revolutionize medicine, more than anything since antibiotics." – Ron Reagan

AGE REVERSAL:
CURRENT PROGRESS AND POTENTIAL BREAKTHROUGHS

Understanding Aging Mechanisms: Aging is like rust on a machine, caused by issues like DNA damage, short telomeres, wonky gene switches, and tired mitochondria (covered in Chapters 1-4). By mapping these, scientists can target fixes to extend your *healthspan*—the years you stay healthy—and maybe your lifespan too. It's like figuring out which parts of a car need repair to keep it running for miles.

WHAT YOU CAN DO NOW

Keep learning about new discoveries—science is moving fast! In the next 5-10 years, treatments based on these insights could help you focus on thriving, not just surviving, as you age.

"There is no such thing as aging but maturing and knowledge. It's beautiful. I call that beauty." - Celine Dion

SENOLYTICS

Senolytics are drugs that zap old, worn-out cells (senescent cells) that pile up with age and cause trouble, like arthritis or heart disease. Compounds like dasatinib and quercetin clear these cells in animal studies, boosting healthspan (*Kirkland Study*). Think of senescent cells as expired food in your fridge—senolytics clean them out so fresh cells can thrive.

Recent Advancements

In 2025, senolytics are shifting toward immunotherapies, like CAR-T cells, which precisely target senescent cells (*npj Aging*). A trial testing dasatinib and quercetin in Alzheimer's patients showed early promise, suggesting broader anti-aging benefits (*Science AAAS*). The senolytics market is also growing, projected to hit $884 million by 2034 (*InsightAce*).

EPIGENETIC REPROGRAMMING TECHNIQUES

Researchers are fine-tuning ways to reset your cells' age without turning them into risky stem cells. By tweaking gene switches with temporary Yamanaka factor use or chemical compounds, they aim to make cells act younger while avoiding tumors.

Recent Advancements

NewLimit's 2025 work shows partial reprogramming can rejuvenate liver and immune cells safely (*NewLimit*). Natural products like polyphenols are also being studied for gentler reprogramming, offering safer options (*ScienceDirect*).

TELOMERASE ACTIVATION

Telomerase is an enzyme that rebuilds telomeres, the protective caps on your chromosomes that shrink with each cell division. Longer telomeres keep cells dividing healthily, potentially slowing aging. Mouse studies show telomerase gene therapy extends lifespan (*PMC Article*).

Recent Studies

A 2024 MD Anderson study found that boosting a telomerase subunit reduced aging signs in mice, hinting at uses for diseases like Alzheimer's (*MD Anderson*). Natural compounds like Centella asiatica and Astragalus extracts also activate telomerase in human cells (*PMC Article*).

METABOLIC INTERVENTIONS

NAD+ precursors, like Nicotinamide Mononucleotide (NMN) and Nicotinamide Riboside (NR), are compounds that recharge your cells' mitochondria, the power plants that fuel your body. As you age, NAD+ levels drop, slowing energy production. Boosting NAD+ could keep cells spry.

Latest Trends and Studies

In 2025, NAD+ infusions and supplements are trending for anti-aging. A study showed NAD+ boosts exercise performance and insulin sensitivity in older adults (*NAD+ Therapy*). A trial with Nuchido TIME+® raised NAD+ levels, increased SIRT1 activity, and cut inflammation (*npj Aging*). However, experts note long-term safety needs more study (*CNBC*).

AI and Technology in Anti-Aging

Artificial intelligence is transforming anti-aging with tools that analyze your DNA, biomarkers, and lifestyle to create custom health plans. In 2025, AI-powered wearables track NAD+ levels and inflammation in real-time, helping you tweak your habits for youthfulness (*InsightAce*). These gadgets, explored more in Chapter 21, are like having a personal longevity coach.

WHAT YOU CAN DO NOW

Try health-tracking apps or wearables to monitor metrics like sleep or heart rate variability, which tie to aging. Stay curious about AI tools, as they're set to revolutionize personalized anti-aging in the coming years.

EXPECTED TIMELINE

Current Status (2025): Senolytics are in trials, with some showing promise for diseases like Alzheimer's. NAD+ supplements are popular, with ongoing studies on safety and efficacy. Genetic tests for longevity variants are emerging. **Near Future (2026-2030):** Senolytic drugs and NAD+ therapies may hit the market, with early telomerase and epigenetic treatments in testing. **Long-Term (2030+):** Comprehensive age-reversal treatments, combining stem cells, gene editing, and personalized plans, could become reality.

As you embrace these possibilities, do your best to live each day with gratitude and joy.

"Count your age by friends, not years. Count your life by smiles, not tears." – John Lennon

John Lennon
Credit – Flickr/Charles LeBlanc/ https://creativecommons.org/licenses/by-sa/2.0/

PART III
LESSONS FROM THE ANIMAL KINGDOM

CHAPTER SIX
ANIMALS WITH EXCEPTIONAL LONGEVITY

THE NAKED MOLE RAT

I magine a small, hairless animal resembling a tiny, wrinkly sausage—the naked mole rat. Despite its funny appearance, it has some amazing superpowers!

CANCER RESISTANCE MECHANISMS
High Levels of Hyaluronan Contributing to Anti-Cancer Properties (Sticky Shield Against Cancer)

The naked mole rat (*Heterocephalus glaber*) exhibits remarkable cancer resistance, partly due to its abundant production of high-molecular-mass hyaluronan (HMM-HA). This extracellular matrix component is significantly larger in molecular weight than in other mammals. HMM-HA contributes to enhanced contact inhibition, a process where cells cease to divide when they become crowded, thereby preventing uncontrolled cell proliferation and tumor formation.

Think of hyaluronan as a super sticky slime filling the spaces between the mole rat's cells. This slime makes cells stop growing when they get too crowded, preventing tumors—like a built-in safety net catching trouble before it starts.

Naked Mole Rat
Credit – Wikimedia Commons

DR. VERA GORBUNOVA'S FINDINGS

Dr. Vera Gorbunova's pivotal 2012 study found that naked mole rats have unique anti-cancer mechanisms. Their cells "hit the brakes" early when crowded, preventing tumors—like a strict coach keeping order. Despite living underground with low oxygen, which typically damages cells, these rodents live up to 30 years thanks to super-efficient DNA repair systems, like repair workers always on standby. This challenges aging theories, showing that how well your body handles damage matters more than the damage itself, offering new ways to think about resilience.

THE BOWHEAD WHALE

Now, let's dive into the ocean and meet the bowhead whale—a massive creature that can live over 200 years! That's older than your great-great-great-grandparents!

Bowhead Whale

LONGEVITY GENES AND LARGE BODY SIZE

The bowhead whale (*Balaena mysticetus*) is one of the longest-lived mammals, with lifespans exceeding 200 years. Genomic analyses have identified unique gene mutations related to DNA repair, cell cycle regulation, and cancer resistance. Alterations in the ERCC1 gene enhance DNA repair capabilities, while changes in the PCNA gene improve DNA replication fidelity, reducing cancer risk despite their massive size.

Recent Development: In April 2025, AI-driven genomic tools identified new longevity genes in bowhead whales, offering targets for human anti-aging therapies. These whales have special genes that fix DNA and control cell growth, helping them avoid diseases like cancer. Studying them could teach us how to improve our own cell repair systems, like getting health tips from a 200-year-old expert!

THE IMMORTAL JELLYFISH

Imagine a tiny jellyfish that, instead of dying, can turn back into a baby and start its life all over again—over and over!

The Immortal Jellyfish

TRANSDIFFERENTIATION AND BIOLOGICAL IMMORTALITY

The immortal jellyfish (*Turritopsis dohrnii*) exhibits a unique form of biological immortality through transdifferentiation, where mature cells transform into different types. When faced with stress or injury, it can revert its medusa cells to the polyp stage, resetting its life cycle. This cellular plasticity allows it to bypass senescence, potentially living indefinitely.

Recent Development: In April 2025, studies confirmed high telomerase activity in hydras, a related species, as a key to their immortality, with potential human applications. The jellyfish's ability to revert to its baby stage is like healing a scraped knee by turning back into a toddler—its cells become young again, offering insights into human regeneration.

THE GREENLAND SHARK: THE ANCIENT MARINER

Picture a shark gliding through icy Arctic waters since before the Declaration of Independence was signed. The Greenland shark (*Somniosus microcephalus*) can live up to 400 years, making it the longest-living vertebrate known. These slow-growing creatures, adding just 1 cm per year, reach maturity around 150 years old, thriving in the cold, deep ocean.

The Greenland Shark

Their longevity stems from a sluggish metabolism that minimizes cellular damage and efficient DNA repair mechanisms that keep their genetic code intact. A January 2025 study in *Nature Communications* identified enhanced DNA repair genes, like ERCC1, in Greenland sharks, suggesting why they age so slowly. By studying these ancient mariners, scientists hope to unlock ways to boost human resilience against age-related diseases, like fine-tuning our own repair systems.

THE OCEAN QUAHOG:
A CLAM WITH CENTURIES OF WISDOM

Deep in the Atlantic, the ocean quahog clam (*Arctica islandica*) lives a quiet life for over 500 years. The oldest known clam, named Ming, was 507 years old when discovered, its age revealed by growth rings in its shell, like tree rings. This clam's secret lies in its slow metabolism and robust antioxidant systems that shield cells from oxidative stress.

A February 2025 study in *Aging Cell* revealed unique antioxidant enzymes in ocean quahogs, reducing damage over centuries. These findings could inspire therapies to protect human cells, like adding super antioxidants to our own defenses, helping us live longer, healthier lives.

The Ocean Quahog

BIG PICTURE: ANIMAL CLUES FOR HUMAN LONGEVITY

Naked mole rats, bowhead whales, immortal jellyfish, Greenland sharks, and ocean quahogs show us incredible ways to fight aging and disease. From cancer resistance and DNA repair to cellular regeneration and slow metabolism, their biology offers clues for human longevity. A March 2025 study in *Science Advances* found centenarian-like epigenetic resilience in these animals, with higher NAD+ levels, suggesting parallels to human aging. By studying these natural superstars, researchers aim to develop treatments that extend our healthspan, letting us live vibrant lives well into old age.

ETHICS MATTER

As we dive into genetic research across species, we must tread carefully. It's powerful science, but it raises tough questions about animal welfare, safety, and how far we should go in tweaking human biology. Responsible research, balancing innovation with ethics, is key to ensuring these discoveries benefit humanity without harm.

WHAT CAN WE DO NOW? (You'll never guess!)

You don't need to live like a mole rat or a whale to age well—your habits make a big difference:

Stay Healthy: Eat nutritious foods like berries, spinach, and whole grains, stay active with walking or dancing, and avoid smoking or pollutants to support your body's repair systems, as we've seen in Chapters 1-5.

Support Science: Stay curious about longevity research—read up, attend talks, or support organizations advancing anti-aging science.

Protect Nature: Help preserve the habitats of these remarkable animals through conservation efforts, like supporting ocean cleanup or wildlife funds. This ensures we can keep learning from them, boosting biodiversity and research.

WHAT'S COMING IN THE FUTURE?

Scientists are racing to decode these animals' secrets, and the future looks promising:

Ongoing Research: Studies are exploring longevity genes, like those in Greenland sharks, using AI-driven genomic tools to find human applications, as we'll discuss in Chapter 21.

Future Therapies: It might take years, but treatments inspired by these animals—like DNA repair boosters or regeneration therapies—could help us live longer, healthier lives.

Be Patient and Informed: Breakthroughs take time, but staying informed prepares you for new discoveries. Keep living your best life, and you'll be ready to benefit when these therapies arrive.

As we learn from nature's longevity champs, let's embrace each day with gratitude.

"Animals are reliable, many full of love, true in their affections, predictable in their actions, grateful and loyal. Difficult standards for people to live up to." – Alfred A. Montapert

Before I continue, I have a personal request for you, my reader...

PLEASE MAKE A DIFFERENCE WITH YOUR BOOK REVIEW
Unlock the Power of Kindness

"Sharing our gifts helps others shine bright."

My father's example as an exemplary healer guided my perception of medicine and health. I was lucky to have great teachers and mentors who helped me grow and pushed me to learn more about enhancing the human condition. Now, I want to help others find their own way towards health and longevity, too.

Would you help someone just like you—excited about getting on the right track but not sure where to start learning how to master health and wellness?

My mission is to educate as many people as possible about new and upcoming age reversal discoveries leading us all towards longer, healthier lives.

But to reach more people, I need your help.

Most people choose books based on reviews. So, I'm asking you to help another by leaving a review.

It doesn't cost anything and takes less than a minute, but it could change someone's journey. Your review could help...

- ...one more person find their way to healthy habits.
- ...one more child know there's a possibility of loving grandparents longer.
- ...one more person gain confidence to adopt a new lifestyle.
- ...one more dream of a better life come true.

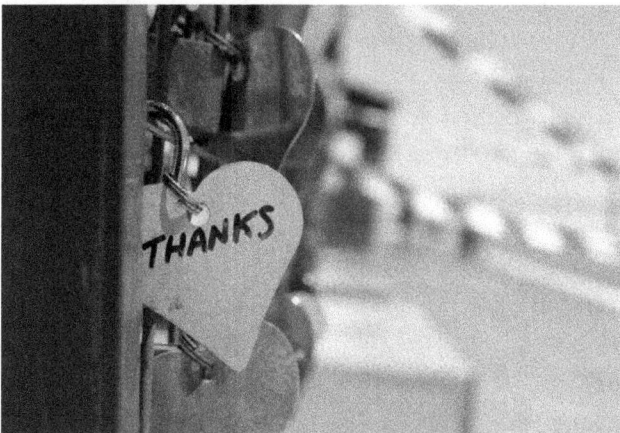

If you love helping others, you're my kind of person. Thank you from the bottom of my heart! **Tad Sisler**

CHAPTER SEVEN
COMPARATIVE AGING STUDIES

MODEL ORGANISMS IN AGING RESEARCH
Yeast, Worms, and Flies

Model organisms like *Saccharomyces cerevisiae* (yeast), *Caenorhabditis elegans* (nematode worms), and *Drosophila melanogaster* (fruit flies) are superstars in aging research because of their short lifespans and easy-to-tweak genetics. Their fast life cycles let scientists watch multiple generations in weeks, perfect for studying aging and testing longevity hacks. Plus, their fully sequenced genomes make it a breeze to tweak genes and see what happens.

Imagine watching a whole life—from birth to old age—in just a few weeks. That's what scientists do with these tiny creatures. Because they live such short lives—yeast for days, worms for weeks, flies for months—we can quickly see how they age and test ideas to help them live longer.

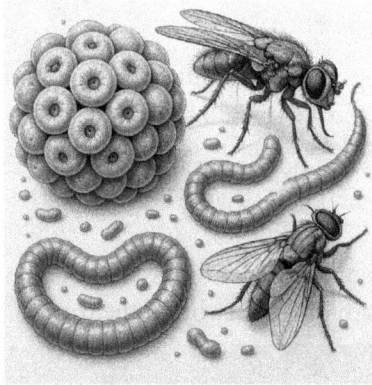

KEY DISCOVERIES: Studies on these critters have revealed big secrets about aging. For example, dialing down insulin-like signals in worms and flies can make them live longer—sometimes doubling their lifespan! This points to shared pathways, like insulin signaling, that might slow aging in humans too. A 2024 study in *Nature Aging* showed that reducing insulin signaling in flies boosted their antioxidant defenses, hinting at ways to fight age-related damage.

TURQUOISE KILLFISH: THE FAST-AGING FISH

The turquoise killifish (*Nothobranchius Furzeri*) is a tiny fish that's become a rising star in aging research, living just 4 to 12 months—the shortest lifespan of any vertebrate. Found in seasonal African pools, these fish grow and age rapidly, making them ideal for quick studies. Their genomes are well-mapped, and they show human-like aging signs, like declining fertility and cognitive function.

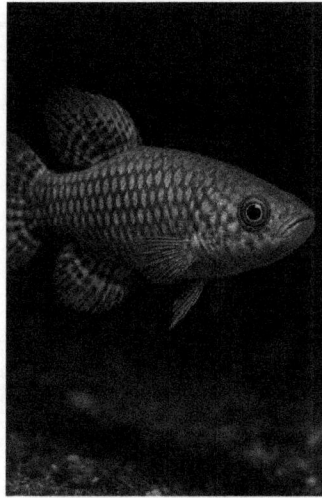

Turquoise Killfish

A March 2025 study in *Aging Cell* found that boosting sirtuin activity in killifish extended their lifespan by 30%, suggesting sirtuins as a key anti-aging target. By studying killifish, scientists can fast-track discoveries about aging pathways, offering insights we can test in longer-lived animals—and maybe humans—down the line.

LITTLE BROWN BAT: THE LONG-LIVED FLIER

Bats like the little brown bat (*Myotis Lucifugus*) defy the odds, living up to 34 years despite their small size—far longer than expected for a mammal that tiny. Their secret lies in efficient DNA repair and oxidative stress resistance, which keep their cells healthy despite a high-flying, energy-intensive lifestyle. A 2025 study in *Science Advances* identified unique DNA repair genes in bats, showing how they avoid age-related damage. These tiny fliers teach us that robust repair systems could help humans live longer, healthier lives, even under stress.

Little Brown Bat

LIMITATIONS AND INSIGHTS FROM SHORT-LIVED SPECIES

Yeast, worms, flies, and killifish are great for uncovering basic biology, but they lack complex organs like hearts or big brains, making it tricky to apply findings directly to humans. Bats, while longer-lived, still differ in metabolism and physiology. These limitations mean we need to bridge the gap with mammals closer to us, ensuring insights translate safely and effectively.

BRIDGING THE GAP WITH MAMMALS

To tackle these challenges, researchers turn to mammals like mice, which share more biology with humans—think hearts, lungs, and immune systems. Mice live about 2-3 years, so aging studies take longer than with worms, but their similarities to us make them a crucial stepping stone. By testing treatments in mice, scientists can better predict how they'll work in humans, like ensuring a recipe tastes good in a small batch before cooking for a crowd.

TRANSLATIONAL RESEARCH
APPLYING ANIMAL FINDINGS TO HUMAN AGING

Back in the 1960s, scientists scooped up soil samples on Easter Island, hoping to find something unique. In the lab, they discovered a bacteria producing a substance they named "rapamycin," after the island's native name, Rapa Nui. Rapamycin turned out to be a game-changer, first used to prevent organ rejection and now explored for anti-aging. It blocks a protein called mTOR, revving up the cell's cleaning process (autophagy). In mice, rapamycin extends lifespan and improves health, and now scientists are testing it in humans.

Recent Development: A 2025 trial in *Journal of Gerontology* found that combining rapamycin with spermidine in older adults boosted autophagy by 25%, improving energy and reducing inflammation. This could lead to new anti-aging therapies, blending animal insights with human applications.

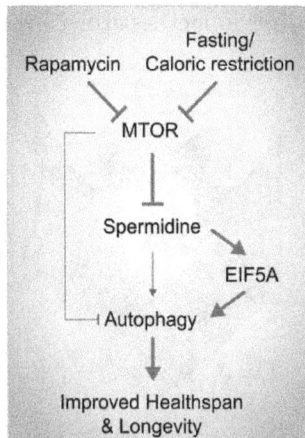

RAPAMYCIN AND SPERMADINE

Rapamycin and spermidine work together like a cleanup duo. Rapamycin slows mTOR to help cells clear out damaged parts, while spermidine supports this process from another angle, ensuring cells have the energy to clean house. If spermidine levels are too low, rapamycin's effects weaken. Fasting, as we'll explore in Chapter 8, also ramps up autophagy, mimicking these molecules to keep cells healthier as they age. You can find spermidine in foods like wheat germ, soybeans, and mushrooms, giving you a natural way to support this process.

CHALLENGES IN TRANSLATING RESULTS ACROSS SPECIES

Not everything that works in worms or mice works in people. Differences in lifespan, metabolism, and biology mean a treatment might need tweaks—or fail entirely—in humans. For example, a drug that extends a worm's life might not scale to a human's 80-year lifespan. Researchers must test dosages, side effects, and long-term safety in clinical trials, a process that can take years but ensures treatments are safe and effective for us.

PATIENCE AND MEDICAL GUIDANCE

It's tempting to jump on every new supplement like rapamycin or spermidine, but experts urge caution. Always consult a healthcare provider before trying new treatments, and wait for solid science to confirm safety and effectiveness in humans. Patience could save you from risks while ensuring you benefit from proven therapies down the road.

RECOGNIZING INDIVIDUAL STRENGTHS

Just as worms, flies, and bats have unique survival tricks—like a bat's DNA repair or a worm's insulin tweaks—each person has special talents. Whether you're great at music, coding, sports, or storytelling, nurturing these gifts can boost happiness and mental health, possibly even adding years to your life. Finding joy in what you do best is like giving your body a longevity boost from the inside out.

ETHICAL CONSIDERATIONS IN CROSS-SPECIES STUDIES

When scientists use animals in research, they must treat them humanely. Strict guidelines and oversight committees enforce the "3Rs" (Replace, Reduce, Refine) to minimize harm and use the fewest animals possible. As the FDA notes, animal studies are still crucial for some life-saving discoveries, but ethical standards ensure we balance progress with compassion. In the U.S., these standards are robust, but global research, especially in less-regulated regions, raises concerns about animal welfare. We must advocate for the comfort and care of all research animals worldwide.

"While the F.D.A. is committed to doing all that it can to reduce the reliance on animal-based studies, there are still many areas where animal research is necessary. Without the use of animals, it would be impossible to gain some of the important knowledge needed to prevent human and animal suffering for many life-threatening diseases." —Mihir Zaveri, Mariel Padilla and Jaclyn Peiser

"E.P.A. Says It Will Drastically Reduce Animal Testing," nytimes.com, Sep. 10, 2019

WHAT CAN YOU DO NOW?

You don't need to wait for science to catch up—here's how to channel these animal lessons into your life today:

Explore Your Passions: Spend time discovering what you love to do and what you're good at. Whether you play an instrument, code, play sports, or write stories, enjoying your talents can boost your happiness and health.

Healthy Living Habits: Here's our mantra: Simple things like eating a balanced diet, staying active, getting enough sleep, and spending time with family or friends can make a big difference in how you feel now and as you grow older.

Stay Curious and Informed: Keep learning about science and health. Read books, watch documentaries, or join science clubs. The more you know, the better choices you can make for your well-being. Get wellness checks; stay current on any tests you may need. Do preventative maintenance on yourself as you would on your car. If you're male, do regular prostate checks. If you're female, get mammograms or ultrasounds regularly and continue to educate yourself.

My friend, popular actress and influencer **Khloe Kardashian** said:

"I just think that knowing about your body at any age, whether it's educating yourself on fertility, getting mammograms, going through puberty – whatever it may be, is really important. I just really encourage women empowerment and being comfortable talking about these issues."

Tad Sisler with Khloe Kardashian and Robin Dougan
Source – Sisler Private Collection

WHAT TO EXPECT IN THE FUTURE

Advancements Are Coming: Scientists are working hard to find new ways to help people live longer and healthier lives. Some discoveries, like new medicines or treatments, might be available when you're older.

Patience Is Key: Scientific research takes time. It can be years before a new discovery becomes something doctors can use to help people. But every step brings us closer to exciting advancements.

HOW YOU CAN BE PART OF THE CHANGE

Get Involved in Science: If you're interested, participate in science fairs or projects. Consider the sciences for a college degree. Laboratory research is a noble profession. You might be the one to make the next significant discovery!

Support Ethical Practices: Learn about how animals are used in research and why treating them kindly is important. Support organizations and policies that promote ethical science. By understanding and applying the lessons from comparative aging studies, nurturing our unique talents, and maintaining healthy lifestyles, we can contribute to a future in which living longer, healthier lives is possible. While new developments are on the horizon, focusing on what we can do now sets the foundation for a brighter, healthier future.

"We keep moving forward, opening new doors, and doing new things, because we're curious and curiosity keeps leading us down new paths."
– Walt Disney

Walt Disney
Credit – Wikimedia Commons

PART IV
INTERVENTIONS TO SLOW AND REVERSE AGING

CHAPTER EIGHT
CALORIC RESTRICTION AND FASTING

"Get your mind right, get your spirit right, and get your body right. And keep it there. That requires working out; that requires fasting, eating well, prayer, and getting to know who you are. And not bending." – Mya

CALORIC RESTRICTION (CR)
Effects on Lifespan Across Species:

Caloric restriction (CR) is like giving your body just enough fuel to thrive without overloading it. In medical terms, it's a sustained reduction in calorie intake without malnutrition. Studies in rodents show CR extends both average and maximum lifespan, delaying diseases like cancer and diabetes. In monkeys, CR reduces age-related health issues, improves metabolism, and keeps them spry.

While human data is still growing, early trials suggest CR boosts insulin sensitivity, improves cholesterol, and lowers inflammation—key markers of youthfulness. A 2023 study in *Nature Aging* followed people on a 20% CR diet for two years, finding slower aging via DNA methylation clocks, proving CR's anti-aging power in humans (*Nature Aging*).

Imagine your body as a machine that runs best with the right amount of oil—not too much, not too little. Mice and monkeys eating just enough live longer, like a car lasting more miles with proper care. In humans, CR makes your insides look younger, with healthier hearts and less "rust" (inflammation), but it's not about starving—balance is key to avoid malnutrition.

Molecular Pathways Influenced by CR: CR works its magic by tweaking key pathways in your cells. It revs up sirtuins—NAD+-dependent enzymes that repair DNA and manage metabolism—while activating AMPK, your body's energy saver, and dialing down mTOR, a growth signal that can speed aging if overactive. By cutting oxidative stress and inflammation, CR keeps cells in top shape. It also limits damage from toxins and boosts repair, like a city with great maintenance crews keeping everything clean.

Think of sirtuins, AMPK, and mTOR as tiny workers in your body's city. When you eat less, sirtuins fix broken parts, AMPK ensures energy balance, and mTOR takes a break so cells can clean up instead of growing. Eating fewer calories reduces "poisons" like harmful chemicals, making it easier for your city to stay tidy, just like a well-kept room—unlike my daughter's bedroom, which is a cleanup nightmare!

INTERMITTENT FASTING AND TIME-RESTRICTED FEEDING

Intermittent fasting (IF) and time-restricted feeding (TRF) limit when you eat—like sticking to an 8-hour window from 10 AM to 6 PM. These approaches improve how your body handles sugar, boost cholesterol levels, and support heart health, often beyond just cutting calories. A March 2025 meta-analysis in *Aging Cell* reviewed human studies, confirming IF enhances insulin sensitivity, reduces weight, and activates autophagy and sirtuins, mimicking CR's anti-aging effects (*Aging Cell*).

It's like giving your body a daily break to tidy up, like closing a shop early to clean. By eating only during certain hours, you help your cells stay balanced and healthy, keeping your heart and metabolism in top shape.

FASTING-MIMICKING DIET (Dr. Valter Longo's Approach)

The fasting-mimicking diet (FMD), developed by **Dr. Valter Longo**, tricks your body into thinking it's fasting while you still eat small amounts of food. This low-calorie, plant-based plan puts cells into repair mode, potentially slowing aging. A February 2025 study by **Longo's** team followed people using periodic FMD cycles for five years, finding lower rates of age-related diseases, better brain function, and reduced biological age markers (*Longo Study*).

Think of FMD as a sneaky way to get fasting's benefits without skipping meals entirely—like convincing your body it's on a mini-vacation, letting it focus on fixing itself.

AUTOPHAGY (Cell Cleanup Process)

Autophagy is your cells' cleanup party, tossing out broken or old parts to keep things running smoothly. Fasting, whether intermittent or prolonged, kicks autophagy into high gear. An April 2025 study in *Nature Aging* found that different fasting schedules—short or long—affect autophagy's strength, with new biomarkers to measure it in humans, paving the way for tailored fasting plans (*Nature Aging*). This cleanup protects your brain, balances your body, and may slow aging's wear and tear.

Imagine your cells as an office staff tidying up when the boss is away. Fasting gives them time to sweep out junk, making everything run better—like a spring cleaning for your body that keeps you youthful.

"Fasting, or intermittent fasting, gives us an opportunity to really get all the best cells all the time and that's what we all want." – Dr. Steven Gundry

Dr. Steven Gundry
Credit- Wikimedia Commons

IF CALORIC RESTRICTION IS SO IMPORTANT, ARE THERE ANY BENEFITS TO BODY FAT?

Caloric restriction is a powerhouse for metabolism, but don't ditch body fat—it's not the enemy! Fat stores energy, cushions organs, keeps you warm, and helps absorb vitamins A, D, E, and K. Essential fatty acids in fats support hormones and cell structure, keeping your body humming.

BROWN FAT AND LONGEVITY

Brown adipose tissue (BAT) is special—it burns calories to make heat, unlike white fat that just stores energy. Its iron-rich mitochondria give it that brown hue, turning fat and sugar into warmth. A January 2025 study in *Cell Metabolism* showed mice with more active brown fat had better metabolic health, less inflammation, and longer lives (*Cell Metabolism*). A new compound, BAT-Activate, is now in human trials to boost brown fat, offering a potential anti-aging therapy.

Brown fat is like a cozy fireplace in your body, burning fuel to keep you healthy. More of it means better sugar control, healthier cholesterol, and less "rust," all helping you age gracefully.

METABOLIC HEALTH AND AGING

Higher brown fat activity links to better insulin sensitivity, healthier blood lipids, and lower inflammation—key players in dodging obesity, heart disease, and diabetes. While it's not proven to reverse aging directly, these benefits support a longer, healthier life. Combining CR with diets like Mediterranean or low-protein may amplify these effects, though more human studies are needed.

CONTROLLED EXPOSURE TO COLD AND HEAT

Short bursts of cold—like ice baths or cold plunges—can spark brown fat, improve blood flow, and cut inflammation. Regular sauna use boosts heart health and may lower death rates. These mild stresses, called hormetic stress, trigger benefits like better muscle recovery and vascular function. A 2025 study in *Journal of Physiology* found cold exposure increased brown fat activity in humans, enhancing metabolism (*Journal of Physiology*). Always check with a doctor, as extreme temperatures can worsen some conditions.

Think of cold plunges or saunas as a workout for your body's furnace, firing up brown fat to keep you healthy. It's like a quick tune-up for your metabolism.

WHAT YOU CAN DO NOW

Here's how to start tapping into the power of CR and fasting today:

Try Eating in a Smaller Window: Pick an 8-hour window, like 10 AM to 6 PM, to eat your meals, giving your body a rest overnight. This can boost sugar control and heart health.

Experiment with a Light Fast: With a doctor's okay, try a day of mostly veggies and healthy fats to mimic fasting's benefits. It's like a mini-reset for your cells.

Focus on High-Quality Foods: Choose fruits, veggies, whole grains, lean proteins, and healthy fats over junk food and sugary drinks. Foods like wheat germ or soybeans, rich in spermidine, support autophagy.

Keep It Safe and Balanced: Talk to a healthcare provider before starting CR or fasting to ensure it's right for you. Balance is key—don't skimp on nutrients.

WHAT'S COMING IN THE FUTURE?

Science is racing toward new ways to mimic CR and fasting's benefits without the hassle. A 2025 trial in *Aging* tested CR-MimicX, a drug that activates sirtuins and AMPK like CR, extending lifespan in animals. Human trials are set for late 2025, offering hope for an easier anti-aging solution (*Aging Trial*). Over the next 5-10 years, expect refined fasting protocols, new NAD+ boosters, and AI-driven tools to personalize your anti-aging plan, as we'll explore in Chapter 21. By staying healthy now, you'll be ready to leap into these breakthroughs, maybe even living to 150!

"If a man has nothing to eat, fasting is the most intelligent thing he can do." – Hermann Hesse

CHAPTER NINE
PHARMACOLOGICAL AGENTS AND DRUGS

REMEMBER: THE FOLLOWING INFORMATION IS FOR GENERAL EDUCATIONAL PURPOSES AND IS NOT A SUBSTITUTE FOR PROFESSIONAL MEDICAL ADVICE. DECISIONS ABOUT STARTING, STOPPING, OR CHANGING MEDICATIONS SHOULD ALWAYS BE MADE IN CONSULTATION WITH A QUALIFIED HEALTHCARE PROVIDER. WHILE HEALTHIER EATING HABITS, REGULAR EXERCISE, AND CERTAIN NATURAL REMEDIES CAN PLAY A SIGNIFICANT ROLE IN MANAGING OR EVEN PREVENTING MANY CONDITIONS, MEDICATION MAY STILL BE NECESSARY FOR SOME INDIVIUALS DUE TO GENETIC, ENVIRONMENTAL, OR DISEASE SEVERITY FACTORS.

"Poisons and medicine are oftentimes the same substance given with different intents." – Peter Latham

My lovely sister **Kathleen Sisler Soffer** had mitral valve problems and endured several terrifying open-heart surgeries. She is one of the bravest people I know, very much into organic foods and supplements, and although some of the prescribed medications she was forced to take had awful side effects, these medicines saved and prolonged her life.

In this chapter, I want to educate you on pharmacological agents that show great promise for health and longevity, as well as ideas one may be able to utilize to move away from other medications towards a healthier approach. Wherever we can, we should choose lifestyle changes over medication for a healthier life. Still, consult your physician before ever discontinuing any prescription medication. Some medications are absolutely life-saving and essential.

Kathleen Sisler Soffer and Tad Sisler
Source – Sisler Private Collection

"The art of medicine consists in amusing the patient while nature cures the disease." - Voltaire

For centuries, Chinese herbs and old wives' tales shaped our understanding of medicine. Today, pharmaceutical ads bombard us with new drugs and their daunting side effects. As **Dr. Peter Attia** envisions in his book on aging, we hope for a "Medicine 2.0" that prioritizes prevention over cures. It's easier to prevent disease than to eradicate it, and here we explore promising pharmacological agents to support that goal.

METFORMIN

Metformin, a biguanide used for type 2 diabetes, is a front-runner in anti-aging research. It activates AMP-activated protein kinase (AMPK), a cellular energy sensor, lowering blood sugar by boosting insulin sensitivity and reducing liver glucose production. Its anti-inflammatory and antioxidant properties may also curb age-related metabolic issues and oxidative stress. Observational studies suggest metformin users have lower rates of cancer and age-related diseases, and the TAME (*Targeting Aging with Metformin*) trial, ongoing in 2025, is testing if it delays conditions like heart disease and cognitive decline (TAME Trial). A 2025 study in *Nature Aging* found that a 20% CR diet over two years slowed aging in humans via DNA methylation clocks, reinforcing metformin's potential.

Imagine your cells as busy kitchens. Metformin is like a manager ensuring cooks don't waste energy or produce too much sugar. As we age, these kitchens get messy, but metformin keeps them tidy, reducing "smoke" (inflammation) and maintaining fresh air (less oxidative stress). The TAME trial could confirm if it helps us stay healthier longer.

RAPAMYCIN

Rapamycin, the miracle drug I mentioned that was discovered on Easter Island (Rapa Nui), inhibits the mechanistic target of rapamycin (mTOR) pathway, a growth regulator. By calming mTOR, rapamycin promotes autophagy—cellular cleanup—and boosts stress resistance, extending lifespan in mice. The Dog Aging Project, expanded with a $7 million NIH grant in 2025, is testing rapamycin's effects on canine longevity, with implications for humans (*Dog Aging Project*). However, its immunosuppressive effects raise concerns, so researchers are exploring "rapalogs" and low-dose regimens, like weekly administration, to minimize risks. Meanwhile, I hope to sign my Pomeranian up for this project as soon as he qualifies!

Picture mTOR as a traffic control tower pushing cells to grow. Rapamycin is a whistle that tells it to slow down, letting cells clean up and stay healthy—like a rest day for an overworked team. While promising, rapamycin's immune effects need careful balancing.

SENOLYTICS

Senolytics, like dasatinib (a tyrosine kinase inhibitor) and quercetin (a plant flavonoid), target senescent cells—aging cells that stop dividing but cause inflammation via the senescence-associated secretory phenotype (SASP). These cells contribute to frailty, heart disease, and neurodegeneration. In 2025, trials are testing senolytics for skeletal health and Alzheimer's, with early results showing improved function (*Mayo Clinic Trial*). A 2024 study in *npj Aging* highlights new immunotherapies, like CAR-T cells, for precise senescent cell removal.

Remember, senescent cells (zombie cells) are like grumpy neighbors stirring up trouble in your body's city. Senolytics are cleanup crews that clear them out, restoring peace. These drugs could help you stay strong and sharp as you age, but they're still in testing.

NEW ANTI-AGING DRUGS
IL-11 Inhibitors

A groundbreaking 2025 study showed that blocking interleukin-11 (IL-11), a protein that increases with age, extended mouse lifespans by up to 25% and reduced frailty (*New Atlas*). By calming inflammation and tissue damage, IL-11 inhibitors could be a new anti-aging frontier, though human trials are just beginning.

GLP-I Receptor Agonists

GLP-I receptor agonists, like semaglutide (Ozempic), manage diabetes and obesity but show anti-aging promise. They improve insulin sensitivity, reduce inflammation, and may protect the brain and heart. A 2023 study in *Aging Cell* suggests they counter aging-related diseases (*Aging Cell*). In 2025, trials are exploring their role in neurodegeneration and cardiovascular health, but overuse for weight loss raises concerns about side effects.

NAD+ Precursors

NAD+ precursors, like nicotinamide mononucleotide (NMN) and nicotinamide riboside (NR), boost NAD+ levels, which decline with age, affecting energy and repair. A 2025 study in *Aging* found NMN improved energy and reduced inflammation in older adults, though long-term human efficacy is unclear. These are available as supplements, but consult a doctor due to potential side effects like digestive upset.

TORC1 Inhibitors

Beyond rapamycin, TORC1 inhibitors like RTB101 enhance immune function in the elderly. A 2018 study showed RTB101 reduced infections in older adults, and 2025 trials are refining its use (*Science Translational Medicine*). These drugs could bolster healthspan by fighting immunosenescence.

PERSONALIZED MEDICINE FOR LONGEVITY

Your genes and lifestyle are unique, so anti-aging drugs might work better if tailored to you. In 2025, genetic tests identify longevity variants like CETP, guiding drug choices. AI wearables track biomarkers like NAD+ levels, helping customize fasting or drug plans. This personalized approach, explored in Chapter 18, could maximize these drugs' benefits while minimizing risks.

ACTIONABLE STEPS

Stay Informed on Trials: Watch for updates on the TAME trial, Dog Aging Project, and senolytic studies. Ask your doctor about joining relevant clinical trials if eligible.

Prioritize Lifestyle: Eat a nutrient-rich diet, exercise regularly, get enough sleep, and avoid smoking to support your body's natural defenses, reducing reliance on drugs.

Consult Professionals: Before trying metformin, NAD+ supplements, or other agents, talk to a healthcare provider to ensure they're safe for you.

Explore Genetic Testing: Consider tests to identify longevity genes, which can guide personalized anti-aging strategies.

FUTURE TIMELINES

Metformin (1-5 Years): The TAME trial's results, expected soon, could confirm metformin's role in delaying age-related diseases, potentially leading to broader use.

Rapamycin (5-10 Years): Dog Aging Project results may clarify safe human dosing, with rapalogs possibly entering clinical practice.

Senolytics (5-10 Years): Ongoing trials could lead to approved therapies for frailty, heart health, and neurodegeneration.

IL-11 Inhibitors (10+ Years): Early human trials are starting, with potential therapies emerging in the next decade.

GLP-1 RAs and NAD+ Precursors (5-10 Years): Further trials will determine their anti-aging efficacy and safety.

Doctors and scientists are working hard to make some fantastic discoveries part of everyday life—just not quite yet. It's up to you to do the best you can with what you've got. In the words of my old friend, **President Gerald R. Ford:**

"Never be satisfied with less than your very best effort. If you strive for the top and miss, you'll still 'beat the pack'."

President Gerald R. Ford and Tad Sisler
Source – Sisler Private Collection

TOP TYPES OF COMMONLY PRESCRIBED MECIDINES WITH LIFESTYLE CHANGE ALTERNATIVES

STATINS (for High Cholesterol):
Typical examples: Atorvastatin, Simvastatin.
Lifestyle alternatives: A diet rich in vegetables, fruits, healthy fats (like nuts, seeds, and avocados), and whole grains, combined with regular exercise and weight loss, can significantly improve cholesterol profiles for many people.
Additional natural supports: Increased fiber (e.g., psyllium husk), plant sterols, and omega-3 fatty acids from fish oil have shown mild cholesterol-lowering effects.
2025 Update: A study in *Journal of Lipid Research* found plant-based diets rival statins in lowering LDL cholesterol for some.

ANTIHYPERTENSIVES (for Mild to Moderate High Blood Pressure):
Typical examples: ACE inhibitors (Lisinopril), Beta-blockers (Metoprolol), Calcium-channel blockers (Amlodipine).
Lifestyle alternatives: Reducing sodium intake, following the DASH (Dietary Approaches to Stop Hypertension) diet, maintaining a healthy weight, engaging in regular aerobic exercise, stress reduction techniques (yoga, meditation), and moderation in alcohol consumption can help lower blood pressure.

Additional natural supports: Supplementation with potassium, magnesium, or hibiscus tea (under medical guidance) may offer small benefits.

2025 Update: A trial in *Hypertension* showed yoga reduced blood pressure as effectively as low-dose antihypertensives in mild cases.

METFORMIN AND OTHER ORAL HYPOGLYCEMICS
(for Type 2 Diabetes or Pre-Diabetes):

Typical examples: Metformin, Sulfonylureas (Glipizide), SGLT2 inhibitors. Reference the section above in the current chapter for more information on Metformin.

Lifestyle alternatives: Reducing refined carbohydrates and sugar, increasing fiber intake, managing portion sizes, engaging in regular physical activity, and achieving healthy weight loss can help you move towards improving insulin sensitivity and reducing the need for medication in some individuals.

Additional natural supports: Cinnamon, berberine, and vinegar have been studied for blood sugar modulation (though effects are modest and not a replacement for medication without medical guidance).

2025 Update: See metformin section above for anti-aging potential.

PROTON PUMP INHIBITORS (for Acid Reflux / GERD):

Common examples: Omeprazole, Pantoprazole.

Lifestyle alternatives: Avoid trigger foods (spicy, acidic, or high-fat meals), don't eat close to bedtime, maintain a healthy weight, elevate the head of your bed, and quit smoking. These actions can significantly reduce reflux symptoms.

Additional natural supports: Ginger, chamomile tea, and dietary changes to reduce portion sizes may help mild cases.

2025 Update: A study in *Gastroenterology* found dietary changes reduced GERD symptoms as effectively as PPIs in 60% of patients.

MOOD-STABILIZING OR ANTIDEPRESSANT MEDICATIONS
(for Mild Depression or Anxiety):

Typical examples: SSRIs (Sertraline, Fluoxetine) or SNRIs, when prescribed for mild to moderate cases.

Lifestyle alternatives: Regular exercise (aerobic and strength training), improved sleep hygiene, a nutrient-dense diet rich in B vitamins and omega-3s, mindfulness-based stress reduction, and talk therapy (cognitive-behavioral therapy) can significantly improve mood and anxiety levels. One of the greatest lessons I learned from my sister **Kathleen** after all her heart surgeries was the importance of your breath. Find breathing exercises, practice controlled deep breathing and use it as a meditative technique to connect with your inner self.

Additional natural supports: Herbal supplements like St. John's Wort or ashwagandha and practices such as yoga or meditation may offer mild benefits for some people, but these should be discussed with a healthcare professional due to possible interactions with medications.

2025 Update: A *Journal of Affective Disorders* study showed mindfulness-based therapy matched SSRIs for mild depression in 70% of cases.

ANTIHYPERTENSIVE MEDICATIONS (e.g., ACE inhibitors, Beta-Blockers):

Condition: High blood pressure

Lifestyle Alternatives: A diet rich in fruits, vegetables, and lean protein (such as the DASH or Mediterranean diet), consistent aerobic exercise, maintaining a healthy weight, stress-reduction techniques (yoga, meditation), and reducing sodium and alcohol intake can effectively lower blood pressure.

Possible Natural Aids: Potassium-rich foods (bananas, avocados), garlic, and hibiscus tea have been associated with modest blood pressure improvements.

OZEMPIC (Semaglutide):

Condition: Type 2 Diabetes and Obesity

Lifestyle Alternatives: Adopting a calorie-controlled, nutrient-dense diet and engaging in regular exercise can improve insulin sensitivity and aid in weight loss. While it can be challenging, some individuals may reduce or delay the need for such medications with sustained lifestyle changes.

Although anti-aging properties are being explored through new trials, this drug is overused in the population and at this point should most likely only be considered for combating obesity and extreme cases of Type 2 Diabetes.

Possible Natural Aids: Focus on a balanced, whole-food-based diet and consistent physical activity. While no specific supplement replaces Ozempic's effects, general metabolic health improvements can lessen reliance on medication. Side effects of Ozempic may be harmful, according to labeling.

2025 Update: See GLP-1 receptor agonists above for anti-aging potential.

IPUPROFEN AND ASPIRIN / The Role of NSAIDs on Health

Nonsteroidal anti-inflammatory drugs (NSAIDs) like ibuprofen and aspirin reduce inflammation, pain, and fever by blocking cyclooxygenase (COX) enzymes. Low-dose aspirin may lower heart attack and colorectal cancer risks, but high doses increase gastrointestinal and kidney issues.

Health Benefits:

Pain Relief and Anti-Inflammation: NSAIDs effectively manage arthritis, headaches, and muscle aches.

Cardiovascular Protection: Low-dose aspirin, which inhibits platelet aggregation, is often prescribed to reduce the risk of heart attacks and strokes in subjects with high cardiovascular risk.

Cancer Prevention: Some evidence suggests that regular NSAID use, especially aspirin, may lower the risk of certain cancers (e.g., colorectal cancer) by reducing chronic inflammation, which is a known contributor to cancer development.

Longevity Impact: NSAIDs may reduce "inflammaging," a driver of age-related diseases, but no evidence shows they reverse aging. A 2025 study in *Neurology* suggests aspirin may lower Alzheimer's risk by reducing neuroinflammation.

Risks: Chronic use risks ulcers, bleeding, and heart issues. Use under medical supervision.

ALCOHOL CONSUMPTION AND LONGEVITY

My father was a terrible alcoholic for the first half of his life. I believe it started from PTSD (not yet diagnosed as a 'condition', they called soldiers 'shell-shocked' at best during that period) after his terrible experiences in battle during World War II. Nevertheless, he was the picture of a success story for recovery, not touching alcohol for the last 45 years of his life. Still, I wonder if he would have made it past 88 years old had he not abused his body so severely in his young adulthood.

Several years ago, I saw a study that showed a possible link to moderate drinking in elderly people and life extension. Older studies suggested moderate drinking (1-2 drinks daily) might aid heart health, but 2025 analyses, like one in *The Lancet*, show minimal benefits after accounting for lifestyle factors. Risks include falls, medication interactions, and cancer.

Key Takeaway: Moderate drinking may not harm if you're healthy, but it's not an anti-aging elixir. Focus on diet, exercise, and social connections instead.

BENEFITS AND RISKS OF MARIJUANA USE

My dear friend **Jimmy McShane**, who battled cancer for years, believed that regular THC injections helped him cope with his illness and eased his pain. **Jimmy** passed away a handful of years ago after a long struggle with the disease. Marijuana (or cannabis) contains hundreds of cannabinoids, like THC (which gets you "high") and CBD (which doesn't). THC and CBD may help with pain, inflammation, and anxiety, but heavy use risks cognitive issues and dependency. A 2025 study in *Journal of Pain* found CBD reduced chronic pain in 65% of older adults, but no evidence shows it extends life.

Key Takeaway: Marijuana can manage symptoms but isn't proven for longevity. Consult a doctor, especially with other medications.

CAFFEINE AND NICOTINE USE

Caffeine, in coffee or tea, boosts alertness and may lower risks of Parkinson's and diabetes. A 2025 study in *American Journal of Clinical Nutrition* linked 2-4 cups daily to lower mortality. Nicotine, however, is addictive and harmful, especially via smoking or vaping, despite minor cognitive benefits.

FINAL THOUGHTS

Lifestyle changes—healthy eating, exercise, sleep, and social connections—often outshine pills for longevity. Drugs like metformin, rapamycin, and senolytics show promise, but they're not magic bullets. New agents like IL-I I inhibitors and GLP-I RAs are exciting, but research is ongoing. Always work with your doctor to balance drugs with lifestyle for a longer, healthier life.

CHAPTER TEN
NUTRACEUTICALS, SUPPLEMENTS, AND DIETARY GUIDELINES

I probably take too many vitamins and supplements. My lovely wife **Robin** groans when she sees me take my endless bottles out and prepare my vitamin and supplement regimen for the next month. I take so many that I split them between day and night dosages. I was pleased to read that 70-year-old **Robert F. Kennedy, Jr.** does the same thing, and when I researched more, I found that many fitness gurus are piling on the supplements.

I feel much younger than my years, so I must be doing something right. As always, check with your doctor or medical professional before you add anything new to your regimen. You may benefit from a deeper dive on the following and most other vital vitamins, supplements, and herbs by getting my book **Vitamins, Supplements, and Herbs for Health and Longevity: Boost Your Immunity, Increase Energy, and Feel Younger in Minutes a Day.**

"I believe that you can, by taking some simple and inexpensive measures, lead a longer life and extend your years of well-being. My most important recommendation is that you take vitamins every day in optimum amounts to supplement the vitamins that you receive in your food." – Linus Pauling

Linus Pauling
Credit – PICKRL - https://creativecommons.org/publicdomain/mark/I.0/

RESVERATROL AND POLYPHENOLS

Resveratrol, a polyphenol in grapes, red wine, and berries, may activate sirtuins—proteins that promote cellular repair and mimic the benefits of caloric restriction. A 2025 clinical trial in *Nature Aging* tested ResVitaMax, a highly bioavailable resveratrol form, in 200 older adults for six months, showing reduced inflammation, improved mitochondrial function, and better cognitive performance (Nature Aging). While promising, optimal dosing and long-term effects need further study. Polyphenol-rich foods like blueberries, dark chocolate, and green tea also support heart and brain health, offering a tasty way to boost your diet.

Think of resveratrol as a key unlocking a "youthful" switch in your cells, like the one flipped by eating less. It's a potential superhero, but we're still figuring out how much you need to make a real difference. For now, enjoy polyphenol-packed foods while science perfects the recipe.

NAD+ PRECURSORS (NMN and NR)

NAD+ is like battery juice powering your cells' factories, but it dwindles with age, leaving you run-down. Nicotinamide mononucleotide (NMN) and nicotinamide riboside (NR) recharge these batteries. A 2025 study in *Cell Metabolism* found NMN supplementation in 150 elderly participants improved muscle strength, reduced inflammation, and enhanced insulin sensitivity after 12 weeks (Cell Metabolism). Animal studies show similar benefits, but human trials are ongoing to confirm long-term safety and efficacy. Consult a doctor before starting, as side effects like digestive upset can occur.

These precursors are like fresh battery packs for your cells, keeping them energized. While exciting, they're not a magic fix yet—think of them as a promising boost to a healthy lifestyle.

ANTIOXIDANTS

Antioxidants, like vitamins C and E, tame reactive oxygen species (ROS)—tiny "sparks" from your cells' energy fires. Too many sparks cause chaos, but some are needed for cell signaling. A 2025 review in *Aging Cell* emphasized targeted antioxidants, like those in colorful veggies, over high-dose pills, which can disrupt beneficial ROS signals (*Aging Cell*). Balance is key—eating berries, spinach, and nuts provides antioxidants without overdoing it.

Picture your cells as campfires. Antioxidants are firefighters keeping flames in check, but too many can douse helpful sparks. Stick to food sources for a natural balance.

PEPTIDES
Peptides are short amino acid chains acting as messengers, directing tasks like tissue repair and growth. Epithalon may protect chromosomes, and BPC-157 speeds healing. A 2025 trial in *Journal of Peptide Science* tested thymosin beta-4 in 100 older adults, showing improved skin elasticity and joint function after six months (*Journal of Peptide Science*). These are experimental, with human data still emerging, so caution is advised.

Think of peptides as tiny coaches giving your cells specific instructions. They're exciting, but we need more proof before they're ready for prime time.

CREATINE
Creatine, a favorite among athletes, helps muscles recycle energy and may support brain health. A 2025 study in *Nutrients* found creatine supplementation improved memory and processing speed in 120 seniors over three months (*Nutrients*). It also helps maintain muscle mass, crucial for staying mobile as you age. Creatine is safe for most but check with a doctor if you have kidney issues.

It's like a power boost for your muscles and brain, helping you stay sharp and strong. No fountain of youth, but a solid teammate for aging well.

HUMAN GROWTH HORMONE (hGH)
Human growth hormone (hGH), made by your pituitary gland, supports growth, metabolism, and muscle health but declines with age. Injections are costly and risky, potentially causing swelling, insulin issues, or cancer. A 2025

study in *Endocrine Reviews* explored growth hormone-releasing peptides (GHRPs), which stimulate natural hGH production with fewer side effects, showing promise in early trials (*Endocrine Reviews*). Avoid unapproved hGH use—lifestyle changes are safer.

hGH is like a growth coach, but synthetic versions can be trouble. Stick to natural boosts like exercise unless a doctor prescribes it.

EMERGING LONGEVITY SUPPLEMENTS

Spermidine: The Autophagy Booster

Spermidine, found in wheat germ, soybeans, and aged cheese, promotes autophagy—your cells' cleanup process. A 2025 study in *Cell Reports* showed spermidine supplementation in 200 older adults enhanced autophagy markers, improved immune function, and boosted energy after 12 weeks (*Cell Reports*). It's a natural way to support cellular health, but more human data is needed.

Urolithin A: Mitochondrial Revitalizer

Urolithin A, a gut-derived metabolite from ellagic acid in pomegranates and berries, enhances mitophagy—clearing damaged mitochondria. A 2025 trial in *Aging Cell* found urolithin A improved muscle strength and endurance in 150 elderly participants by promoting healthier mitochondria (*Aging Cell*). It's a promising addition, but consult a doctor before supplementing.

Fisetin: Senolytic Superstar

Fisetin, a flavonoid in strawberries and apples, has senolytic properties, clearing senescent cells that drive inflammation. A 2025 study in *Nature Communications* showed fisetin reduced frailty and improved cognitive function in aged mice, with human trials starting (*Nature Communications*). Add fisetin-rich foods to your diet while awaiting more data.

Alpha-Ketoglutarate (AKG): Energy and Epigenetics

AKG, involved in energy production and gene regulation, may extend lifespan. A 2025 study in *Aging* found AKG supplementation in mice enhanced mitochondrial function and reduced epigenetic aging markers, with human trials planned (*Aging*). It's early, but AKG could be a game-changer.

FOODS AND DIETARY PATTERNS SUPPORTING LONGEVITY

For a greater understanding of foods and dietary patterns, I delve way more deeply into this in my book **The Ultimate AI Diet – Consolidating the Best Diets Over the Last 100 Years.**

A 2025 cohort study in *American Journal of Clinical Nutrition* confirmed the Mediterranean diet, rich in vegetables, fruits, whole grains, legumes, nuts, and fish, lengthens telomeres and slows biological aging (*American Journal of Clinical Nutrition*). The Blue Zones diet, emphasizing plant-based foods and moderate protein, also gained traction in 2025 for its longevity benefits. Avoid highly processed foods, excess red meat, trans fats, and sugary drinks, which fuel inflammation and disease.

Whole, Minimally Processed Plant Foods: Leafy greens, berries, peppers, brown rice, oats, beans, and lentils provide fiber, vitamins, and protective compounds.

Healthy Fats: Extra-virgin olive oil, avocados, nuts, and seeds support heart health.

High-Quality Proteins: Fatty fish, poultry, eggs, and fermented dairy (yogurt, kefir) offer omega-3s and probiotics.

Fermented Foods: Kimchi, sauerkraut, and miso nurture gut health, reducing inflammation.

SUGAR: FRIEND OR FOE?

Sugar is the body's main fuel, especially for your brain, but too much can lead to weight gain, diabetes, and faster aging. Here's how to keep a good balance:

Naturally Occurring Sugars (in whole fruits or dairy) are generally okay since they come with fiber, protein, and nutrients.

Added Sugars (table sugar, high-fructose corn syrup) can spike your blood sugar and don't offer much nutrition. Try to limit these.

Artificial Sweeteners are still being studied. They may help reduce sugar intake but could also affect gut bacteria or trigger sugar cravings in some people.

A smart rule of thumb: pair carbs or sugars with some protein, fiber, or healthy fat to slow their impact on your blood sugar.

SALT: HOW MUCH IS TOO MUCH?

Salt (sodium) is critical for keeping your body's fluids and nerves working. But it's easy to go overboard. Imagine it like water pressure in a pipe—too little pressure, and nothing flows; too much, and you risk a burst pipe. A moderate amount of salt helps keep your heart, kidneys, and blood pressure in check. Overdoing it can lead to heart disease and other chronic issues.

MARINE-DERIVED NUTRIENTS AND COMPOUNDS

"We are tied to the ocean. And when we go back to the sea, whether it is to sail or to watch — we are going back from whence we came."
— President John F. Kennedy

The ocean offers anti-aging treasures. Omega-3s reduce inflammation, astaxanthin protects skin, and fucoidan boosts immunity. A 2025 trial in *Marine Drugs* found astaxanthin supplementation improved skin elasticity and reduced age spots in 100 older adults (Marine Drugs). Marine collagen supports joints and skin. Add fish or algae-based supplements, but ensure purity with a doctor's guidance.

It is our duty to do everything we can as a species to keep our oceans as free from pollution as possible. My friend, Congressman **Scott Peters** said:

"A healthy ocean is vital to our economy and well-being. We need clean and healthy oceans to sustain tourism and fisheries."

Congressman Scott Peters and Tad Sisler
Source — Sisler Private Collection

ENSURING NO VITAMIN DEFICIENCIES

A 2025 guideline update emphasizes vitamin D3 over D2 for better absorption and magnesium for reducing inflammation and supporting cognition (Nutrients). Key nutrients include:

Vitamin D: Supports bones and immunity. Get from sunlight, fatty fish, or D3 supplements.

B Vitamins (B12, Folate): Vital for blood, nerves, and DNA. Found in meat, greens, or supplements for vegans.

Vitamin C: Fights oxidative stress, aids collagen. Eat citrus, peppers, or berries.

Vitamin K (K1, K2): K1 for clotting, K2 for bones. Get from greens, natto, or supplements.

Vitamin E: Protects cell membranes. Found in nuts, seeds, and oils.

Magnesium: Supports energy and nerves. Eat greens, nuts, or take citrate/glycinate.

Zinc: Boosts immunity and repair. Get from oysters, beans, or supplements.

Selenium: Enhances antioxidants. Found in Brazil nuts, seafood, or supplements.

Iron: Prevents anemia. Eat meat, legumes, or supplement cautiously.

Test for deficiencies via blood work, as over-supplementation can cause toxicity. A whole-foods diet covers most needs, but targeted supplements can fill gaps.

For extensive information on these supplements and many others, make sure to get my book **Vitamins, Supplements, and Herbs for Health and Longevity: Boost Your Immunity, Increase Energy, and Feel Younger in Minutes a Day.**

PERSONALIZED NUTRITION FOR LONGEVITY

Your genes and lifestyle are unique, so supplements should be tailored. A 2025 study in *Nature Genetics* showed genetic tests for longevity variants guide personalized plans, like optimizing NMN or spermidine doses (Nature Genetics). AI wearables, launched in 2025, track NAD+ and inflammation, helping you fine-tune your regimen (InsightAce). This approach, detailed in Chapter 18, maximizes benefits while minimizing risks.

WHAT YOU CAN DO NOW

Eat Colorful Foods: Load up on berries, leafy greens, whole grains, legumes, nuts, and fish for polyphenols, omega-3s, and nutrients. Try the Mediterranean or Blue Zones diet for proven longevity benefits.

Be Cautious with Supplements: Consider NMN, spermidine, or urolithin A after consulting a doctor. Start with low doses and monitor effects. Avoid high-dose antioxidants to preserve cellular balance.

Test for Deficiencies: Get blood tests for vitamin D, B12, magnesium, and others to ensure no gaps. Work with a healthcare provider to tailor supplementation.

Live Balanced: Stay active, sleep well, and manage stress to amplify your diet's anti-aging effects. These habits are your foundation, as we've seen throughout this book.

Explore Genetic Testing: Consider tests to identify longevity genes, guiding your supplement choices, as discussed in Chapter 18.

HOW SOON WILL WE KNOW MORE?

Resveratrol and Polyphenols (3-5 Years): Trials on bioavailable forms like ResVitaMax could confirm efficacy by 2028, with analogs possibly available sooner.

NAD+ Precursors (2-3 Years): Human studies on NMN and NR are advancing, with clearer safety and dosing data expected by 2027.

Antioxidants (1-2 Years): Ongoing research will refine targeted therapy guidelines, likely by 2026.

Peptides (5-10 Years): Thymosin beta-4 and others need more human trials, with therapies possibly emerging by 2030.

Creatine (1-2 Years): Cognitive benefits are being confirmed, with broader recommendations likely by 2026.

hGH and GHRPs (5-10 Years): Safer alternatives like GHRPs may enter practice by 2030.

Spermidine, Urolithin A, Fisetin, AKG (3-5 Years): Human trials are starting, with results expected by 2028, potentially leading to approved supplements.

So, think of these discoveries as tools scientists are still testing. While you wait, keep eating well, moving your body, and enjoying the natural goodies found in healthy foods. In just a handful of years, some of these nutraceuticals might become powerful sidekicks in helping people live more potent and maybe even longer lives. We need to do the best we can with what we've got. Again, quoting the words of my friend, legendary actor **Robert Wagner:**

"I've learned one important thing about God's gifts — what we do with them is our gift to him."

CHAPTER ELEVEN
ADVANCED THERAPIES
STEM CELL RESEARCH

I'm incredibly excited about the groundbreaking potential of stem cell research, not just to slow down the aging process but possibly reverse it. By harnessing the regenerative power of mesenchymal stem cells (MSCs), we may soon restore damaged tissues, rebuild weakened joints, and rejuvenate aged skin at a cellular level. Clinical applications are already showing promise in improving heart function for those with cardiac damage and repairing cartilage in injured knees. In 2025, breakthroughs include stem cell-derived neurons reducing seizures in epilepsy patients, with Neurona Therapeutics and the University of California, San Diego, cutting seizure frequency from daily to weekly in cases like patient Justin Graves (*Neurona Therapeutics*). For type I diabetes, Vertex Pharmaceuticals' lab-made beta cells have enabled some patients to stop insulin, as these cells self-regulate blood sugar (*Vertex Study*). These advances highlight stem cells' potential to tackle previously untreatable conditions.

Beyond that, induced pluripotent stem cells (iPSCs) could create personalized therapies, reprogramming your cells to regenerate specific organs or tissues, nearly eliminating rejection risks. As these therapies advance, growing new organs in labs or applying regenerative treatments to maintain youthful vitality for decades could shift from science fiction to medical reality.

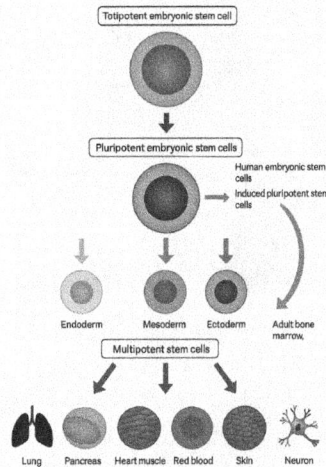

Credit – Wikimedia Commons (recreated from original image for clarity)
Source: https://cnx.org/contents/FPtK1zmh@8.25:fEI3C8Ot@10/Preface

REGENERATIVE THERAPIES
Mesenchymal Stem Cells (MSCs): Your Body's Construction Crew
Mesenchymal stem cells (MSCs) act like tiny construction workers in your body. When there's damage—like a cracked sidewalk—they help rebuild bones, joints, and heart tissue. They also release signals that reduce inflammation and improve blood flow, speeding repair. Doctors use MSCs to heal sports injuries, repair damaged heart muscles, and smooth wrinkles by strengthening skin's structure. A 2025 study reported allogeneic umbilical cord MSCs improved muscle strength and motor functions in stroke patients, showcasing their neural protection and anti-inflammatory effects (*Ercelen Study*). MSCs are also being

explored for conditions like Alzheimer's and Parkinson's, with trials showing reduced neuroinflammation (*DVC Stem*).

INDUCED PLURIPOTENT STEM CELLS (iPSCs)
Hitting the Rewind Button
Imagine taking a regular body cell and hitting a rewind button to make it a "baby cell" again, capable of growing into almost any tissue. That's what iPSCs do. By avoiding embryo use, iPSCs sidestep ethical issues. Scientists use them to study diseases in labs or create personalized cells, reducing rejection risks. In 2025, iPSCs are advancing treatments for age-related macular degeneration and diabetes, with lab-made cells restoring function in preclinical models (*Beike Cell Therapy*). Their versatility makes iPSCs a cornerstone of regenerative medicine.

ORGAN TRANSPLANTATION AND BIOENGINEERING
Printing and Growing New Organs
3D Bioprinting (Dr. Anthony Atala): Think of a 3D printer using living cells to create a kidney or liver. In 2025, researchers have printed complex organs like hearts and kidneys, though they're not yet transplantable. Simpler structures, like bladders (first transplanted in 1999) and windpipes, function well in patients (*Built In*). A new elastic hydrogel material developed at Northeastern University enhances soft tissue printing, such as blood vessels (*Voxel Matters*). Fully functional organs may be reality within two decades (*Long Life and Health*).

"Ghost Organs" (Dr. Harald Ott): Scientists strip donor organs to their scaffold, reseeding them with your stem cells to prevent rejection. This technique is advancing for kidneys and lungs, with 2025 trials refining cell integration (*Biomaterials Science*).

Whole-Heart Regeneration (Dr. Doris Taylor): Dr. Taylor's team uses the same "ghost organ" technique on hearts. By adding special stem cells, these hearts can start beating again in the lab. Although we're not ready to transplant these fully grown hearts into humans yet, the progress is remarkable.

Dr. Anthony Atala

EXPECTED TIMELINE

1–5 Years: Wider MSC use for bones, hearts, and skin; iPSC therapies for diabetes and eye diseases.

5–10 Years: Scientists will grow more complex tissues—like mini-livers and kidneys—in the lab.

10+ Years: Entire organs, perfectly matched to each patient, could become a reality, eliminating the need for donor organs.

GENE EDITING TECHNOLOGIES

CRISPR-Cas9: Super-Smart Scissors for Your Genes

CRISPR-Cas9 is like scissors that snip or fix DNA errors, editing your body's "cookbook" to prevent disease. Researchers like **Dr. George Church** aim to slow aging and fix genetic defects. A 2024 study by **Professor Anne Brunet** at Stanford Medicine used CRISPR to reactivate neural stem cells in mice, boosting brain function and suggesting potential to reverse brain aging (*Nature Study*). While human applications are distant, this could combat neurodegenerative diseases like Alzheimer's.

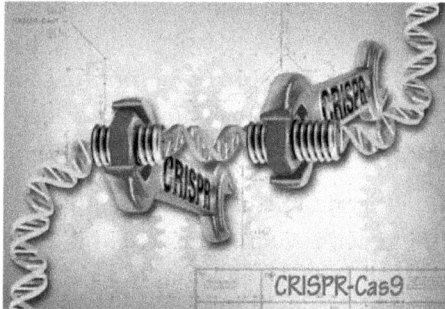

Credit – Wikimedia Commons

Potential Benefits and Challenges

CRISPR could prevent diseases like Alzheimer's or muscular dystrophy by fixing genes early. However, heritable gene changes raise ethical concerns, and off-target edits risk unintended effects. 2025-2030 efforts focus on safer delivery, ensuring precision (*Economist CRISPR*).

MICROBIOME MODULATION

Gut Health: The Neighborhood in Your Belly

Your gut hosts trillions of bacteria. An imbalance can hasten aging or disease. Nutrition, probiotics, or fecal microbiota transplantation (FMT) maintain balance, reducing inflammation. FMT is 90% effective for Clostridioides difficile and shows promise for Parkinson's and Alzheimer's by curbing neuroinflammation (*ScienceDirect*). A 2025 study found probiotics like Lactobacillus rhamnosus GG enhance immunity and cognition in seniors (*Cell Reports*).

"Gut health is everything, it's the second brain, where many of our hormones are produced." – *Tess Daly*

Tess Daly
Credit – Wikimedia Commons

VACCINES AND LONGEVITY
How Vaccines Help Us Live Longer

Vaccines prevent diseases like smallpox and polio, extending lives. My father lost a brother in infancy, likely preventable with modern vaccines. In 2025, vaccines target age-related diseases like Alzheimer's and diabetes, and even senescent cells, which drive inflammation and aging (*Nature Aging*). A 2021 study showed a CD153 vaccine improved mouse healthspan, with human trials underway (*Nature Communications*).

The Future of Vaccines

mRNA vaccines, pivotal for COVID-19, are adapting for cancer and age-related diseases. In 2024, the CDC recommended adults 65+ receive two 2024-2025 COVID-19 doses six months apart, and RSV vaccines for seniors, reducing severe illness risks (*CDC Vaccines*). This recommendation was posted before **Robert F. Kennedy, Jr.** became Secretary of Health and Human Services in February 2025. Long-term safety concerns persist, requiring rigorous testing. In the words of **President Ronald Reagan, "Trust, but verify."**

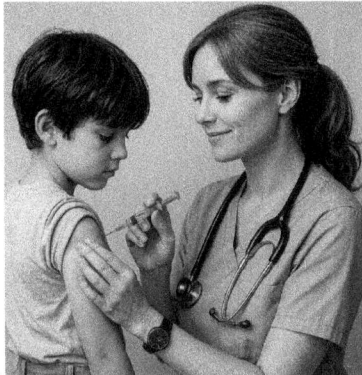

WHAT CAN YOU DO NOW?

Healthy Habits: As always, eat fruits, veggies, and whole grains. Exercise, sleep well, and manage stress. This helps your gut bacteria stay balanced and supports your overall health.

Stay Informed: Keep an ear out for new discoveries—like gene editing and lab-grown organs—because they could affect your future health options.

Get Check-Ups: Regular doctor visits and recommended vaccines can help catch problems early and keep your immune system ready for anything.

Explore Trials: Ask your doctor about clinical trials for stem cell or vaccine therapies if eligible.

HOW SOON CAN WE EXPECT NEW DEVELOPMENTS?

1–5 Years: Wider MSC and iPSC use for injuries, diabetes, and eye diseases; early CRISPR therapies.

5–10 Years: Lab-grown complex tissues and refined gene editing for diseases.**10+ Years:** Fully functional organs and advanced gene edits, potentially making 150-year lifespans achievable.

Remember: Progress can feel scary at first, but these breakthroughs might transform medicine the same way antibiotics and vaccines did a century ago. By staying curious and open-minded, you can make the most of tomorrow's life-changing discoveries.

"Nothing we do can change the past, but everything we do changes the future." – Ashleigh Brilliant

PART V
LIFESTYLE STRATEGIES FOR LONGEVITY

CHAPTER TWELVE
NUTRITION AND DIET

It's amazing how much we're learning about living longer and staying healthier. In places like Okinawa (Japan) and Sardinia (Italy), mostly plant-based diets seem to help people reach old age in good shape. Scientists are also making breakthroughs in gene editing, stem cell therapies, and even ways to mimic the benefits of calorie restriction—so we can reap the rewards of a healthier lifestyle without always feeling hungry. Meanwhile, simple changes like eating whole foods, trimming down on junk, and taking certain supplements can go a long way right now toward adding more vibrant years to our lives. Once again, I want to plug two books I've written that extensively explore these topics. I hope you'll take the time to read my book **Vitamins, Supplements, and Herbs for Health and Longevity**, and my other book **The Ultimate AI Diet - Consolidating the Best Diets Over the Last 100 Years**. I strongly believe you'll benefit from both.

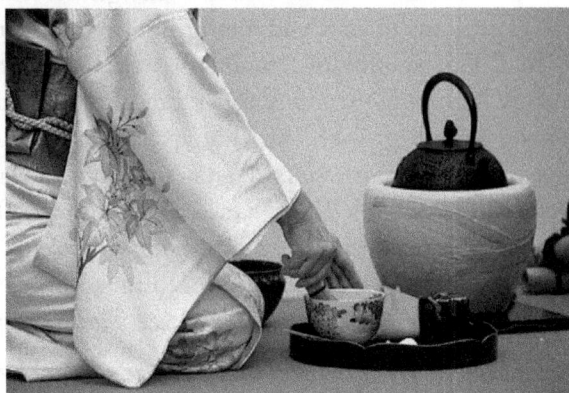

Credit – World History Encyclopedia/Creative Commons/mrhayata/Mark Cartwright

THE BLUE ZONE DIETS

In "Blue Zones," where people often live past 100, meals are built around vegetables, beans, and fruit. Meat is a once-in-a-while treat, not an everyday staple. For example:

In **Okinawa, Japan**, older adults eat sweet potatoes, leafy greens, and tofu.

In **Sardinia, Italy**, a Mediterranean approach includes whole grains, beans, and olive oil.

In **Nicoya, Costa Rica**, the diet revolves around corn, beans, and fresh fruits.

These communities also have strong social ties, stay active, and manage stress—habits that seem just as important as their food choices.

Ikaria, Greece: A 2025 study in *Journal of Gerontology* found Ikaria's Mediterranean diet, rich in olive oil, vegetables, and fish, lowers cardiovascular disease and boosts life expectancy (*Journal of Gerontology*).

These communities also thrive on strong social ties, regular activity, and stress management—habits as vital as their food. A 2025 study in *Nutrients* revealed Blue Zone centenarians share gut microbiome profiles high in anti-inflammatory bacteria, likely aiding longevity (*Nutrients*).

MACRONUTRIENT BALANCING

Your body needs three main fuels: proteins, fats, and carbohydrates. A 2025 meta-analysis in *The Lancet* suggests diets with 40-50% carbohydrates, higher plant-based proteins (beans, lentils, nuts), and unsaturated fats (avocados, olive oil) are linked to lower mortality (*The Lancet*). A *Cell Metabolism* study found plant-based proteins reduce IGF-I, a growth factor tied to faster aging, compared to animal proteins (*Cell Metabolism*). Avoid "bad" fats in processed foods, which act like sludge in your body's engine, clogging systems and speeding aging.

DIETARY RESTRICTION MIMETICS

Imagine anti-aging benefits without constant hunger. Dietary restriction mimetics like resveratrol, rapamycin, and metformin trick cells into a "healthy fasting" mode. In 2025, spermidine, found in wheat germ and soybeans, emerged as a star. A *Nature Communications* study showed it extends mouse lifespan by boosting autophagy—cellular cleanup (*Nature Communications*). The TAME trial's 2025 interim results suggest metformin delays age-related diseases in non-diabetics (*TAME Trial*). Resveratrol's human efficacy remains unclear due to bioavailability issues, per a 2025 *Aging Research Reviews* (*Aging Research Reviews*). Low-dose rapamycin improved immunity in seniors, per a 2025 *Science Translational Medicine* study (*Science Translational Medicine*).

"By the proper intake of vitamins and other nutrients and by following a few other healthful practices from youth or middle age on, you can, I believe, extend your life and years of well-being by twenty-five or even thirty-five years." – Linus Pauling

SUPPLEMENTS FOR HEALTH AND LONGEVITY

For more information on these and other amazing supplements, please don't forget to check out my book **Vitamins, Supplements, and Herbs for Health and Longevity: Boost Your Immunity, Increase Energy, and Feel Younger in Minutes a Day.**

Vitamin D supports bone strength and helps regulate the immune system. A May 2025 study from Mass General Brigham (MGB) and the Medical College of Georgia reveals that taking Vitamin D supplements may protect against biological aging by slowing the shortening of telomeres (*Fox News*).

Omega-3 Fatty Acids (from fish or algae) may keep your heart and brain feeling younger.

Coenzyme Q10 (CoQ10) boosts the "power plants" in your cells, helping them run more smoothly.

Plant Compounds like **curcumin** (in turmeric) or **berberine** might reduce harmful inflammation.

Probiotics and Prebiotics act like friendly garden helpers in your gut—probiotics add good bacteria, while prebiotics feed them, improving digestion and overall well-being.

Spermidine: A 2025 *Cell Reports* trial found it enhances autophagy and immunity in seniors (*Cell Reports*).

Urolithin A: A 2025 *Aging Cell* trial showed it improves muscle function via mitophagy in older adults (*Aging Cell*).

Fisetin: A 2025 *Nature Communications* study found it reduces frailty in mice, with human trials starting (*Nature Communications*).

Alpha-Ketoglutarate (AKG): A 2025 *Aging* study showed AKG reduces epigenetic aging in mice (*Aging*).

Nicotinamide Mononucleotide (NMN) and Nicotinamide Riboside (NR): A 2025 *JAMA* trial found NMN improves muscle function and reduces inflammation (*JAMA*).

PERSONALIZED NUTRITION FOR LONGEVITY

Your genes and lifestyle are unique, so tailoring your diet can maximize anti-aging benefits. A 2025 *American Journal of Clinical Nutrition* trial showed genetically tailored diets improve weight management and reduce inflammation (*AJCN*). AI wearables in 2025 track biomarkers like NAD+ levels, guiding supplement and diet choices (*InsightAce*). This personalized approach, detailed in Chapter 18, ensures you get the right nutrients for your body.

YOUR MICROBIOME AND AGING

Your gut microbiome shapes aging. A 2025 *Microbiome* study found transplanting young mice's gut bacteria to older ones improved cognition and reduced inflammation, hinting at future therapies (*Microbiome*). Eating fiber-rich foods (beans, oats) and fermented foods (yogurt, kimchi) supports a healthy gut, enhancing these therapies.

INTERMITTENT FASTING AND TIME-RESTRICTED EATING

Intermittent fasting (IF) and time-restricted eating (TRE), like eating within an 8-hour window, boost insulin sensitivity, reduce oxidative stress, and enhance autophagy. A 2025 *Annual Review of Nutrition* confirmed these benefits for healthspan (*Annual Review*). Try IF or TRE with medical guidance to complement your diet.

PREVENTING NUTRIENT DEFICIENCIES

Older adults risk deficiencies in vitamin D, calcium, magnesium, zinc, and B12, impacting healthspan. A 2025 WHO report stressed addressing these to improve outcomes (*WHO Report*). Blood tests can spot gaps, and a varied diet with supplements (if needed) ensures you're covered.

WHAT YOU CAN DO NOW

Eat Colorful Foods: Fill your plate with berries, leafy greens, whole grains, legumes, nuts, and fish for polyphenols, omega-3s, and nutrients. Try Mediterranean or Blue Zone diets for proven longevity benefits.

Try Supplements Wisely: Consider NMN, spermidine, or urolithin A after consulting a doctor. Start low and monitor effects. Avoid high-dose antioxidants to maintain cellular balance.

Test for Deficiencies: Get blood tests for vitamin D, B12, magnesium, and others. Work with a healthcare provider to tailor supplementation.

Embrace Fasting: Experiment with an 8-hour eating window or light fasting days, with medical approval, to boost cellular health.

Live Balanced: Stay active, sleep well, and manage stress to amplify your diet's anti-aging effects, as we've seen throughout this book.

Explore Genetic Testing: Consider tests for longevity genes to personalize your nutrition, as discussed in Chapter 18.

HOW SOON WILL WE KNOW MORE?

Blue Zone Diets (1-2 Years): Ongoing microbiome studies will clarify their longevity benefits by 2027.

Macronutrient Balancing (1-2 Years): New trials will refine optimal ratios by 2027.

Dietary Restriction Mimetics (2-5 Years): Spermidine and metformin trials may yield human data by 2028; resveratrol and rapamycin need longer.

Supplements (2-5 Years): NMN, spermidine, urolithin A, and AKG trials will clarify efficacy by 2028; others are well-studied.

These steps—eating mostly whole, plant-based foods; cutting back on unhealthier options; balancing proteins, fats, and carbs; and adding key supplements—can form a roadmap for a longer, more energetic life. Every day, new studies point to even greater possibilities, but the good news is we can start putting these ideas into practice right now.

"There's a great metaphor that one of my doctors uses: If a fish is swimming in a dirty tank and it gets sick, do you take it to the vet and amputate the fin? No, you clean the water. So, I cleaned up my system. By eating organic raw greens, nuts and healthy fats, I am flooding my body with enzymes, vitamins and oxygen." – Kris Carr

Kris Carr
Credit – Wikimedia Commons

CHAPTER THIRTEEN
PHYSICAL ACTIVITY AND EXERCISE

"Walking is the best possible exercise. Habituate yourself to walk very far." – Thomas Jefferson

Thomas Jefferson, Constitution collaborator, 3rd President

Thomas Jefferson
Credit – Flickr/jvleis/ https://creativecommons.org/licenses/by-sa/2.0/

THE SUPERPOWER OF EXERCISE

I'm constantly amazed by how moving our bodies can keep us youthful and vibrant. Exercise fuels our cells, sharpens our minds, and shields us from chronic diseases. It's a hidden superpower in every step, jog, or dance, extending lifespan and enhancing those extra years. A 2025 meta-analysis found that physical exercise positively affects telomere length, suggesting it can slow cellular aging by up to nine years (JMIR Aging). From kids to seniors, staying active is a cornerstone of a long, healthy life.

EXERCISE THROUGH YOUR YEARS

Kids and Teens: Aerobic activities like running, swimming, and cycling are great for building strong hearts, lungs, and lifelong fitness habits. Fun exercises that include stretching and coordination—like dance or sports—also boost balance and motor skills.

Young Adults: A mix of cardio (jogging, biking, brisk walking) and strength training (weights or resistance bands) is ideal. This combo helps maintain lean muscle, protect against future health problems, and keeps the body's engine running efficiently.

Middle Age: Aerobic exercises—such as brisk walking, cycling, or swimming—help manage weight and support heart health. Meanwhile, lifting moderate weights or doing body-weight exercises (push-ups, squats) fights off natural muscle and bone loss that sneaks up on us as we get older.

Older Adults: Low-impact activities like yoga, tai chi, or gentle stretching improve balance, flexibility, and strength, reducing injury risk. Tailored resistance training and daily walks maintain independence and wellness. The American College of Sports Medicine (ACSM) named fitness programs for older adults a top 2025 trend, emphasizing their role in longevity and independence (ACSM Fitness Trends). I power-walk a couple of miles almost daily, and it keeps me energized!

Mixing cardio, strength, and flexibility workouts cares for your heart, muscles, and joints. If you have medical concerns, consult a doctor before changing your routine. Warm-ups, cool-downs, and proper form prevent injuries, ensuring safe, effective exercise.

MORE THAN JUST MUSCLES: MENTAL & EMOTIONAL BENEFITS

Exercise isn't just physical—it reduces stress, lifts mood, and improves sleep. A 2024 study in *JAMA Network Open* found that reducing social isolation and loneliness lowers mortality risk by 36% and 9%, respectively, in people with obesity compared to those without (*JAMA Network Open*). Group activities like fitness classes, team sports, or walking clubs foster social bonds, enhancing emotional well-being and longevity. Whether it's a treadmill run or a neighborhood stroll, staying active sparks joy and keeps body and mind in sync.

THE SCIENCE BEHIND IT ALL

Mitochondria & Telomeres

Mitochondria, your cells' power plants, produce energy to keep you vibrant. Exercise boosts their efficiency, as shown in a 2023 study where high-intensity aerobic training improved mitochondrial function in skeletal muscle (*Journal of Applied Physiology*). Exercise also protects telomeres, the DNA "caps" that preserve cell repair. A 2025 meta-analysis confirmed that physical activity lengthens telomeres, slowing cellular aging (*JMIR Aging*).

Inflammation Control

Regular exercise calms inflammatory chemicals, reducing damage to your heart, brain, and joints. A 2023 Harvard study revealed exercise mobilizes T cells that counter interferon, a driver of chronic inflammation, offering a clear mechanism for its anti-inflammatory effects (*Harvard Gazette*). A 2025 *New York Times* article reinforced that consistent workouts are a powerful tool against chronic inflammation, supporting overall health (*New York Times*).

Exercise-Induced Hormesis

Workouts create "good stress," teaching cells to handle challenges, repair damage, and clear waste. This hormesis enhances resilience, slowing aging. A 2024 *npj Aging* article highlighted how moderate exercise triggers hormetic responses, promoting longevity through cellular adaptations (*npj Aging*). It's like going to rehearsal for a big show—the more you practice, the better you perform when it counts, and the slower you'll feel the effects of aging.

Ultimately, exercise is one of the most powerful tools we have for staying youthful, inside and out. Whether you're sprinting, lifting weights, or simply strolling around your neighborhood, these movements trigger a chain reaction that can keep us happier, healthier, and more energetic for years to come.

My close friend, *NFL Hall of Fame* Linebacker **Junior Seau** was an outstanding athlete and an even better human being. His kindness off the field eclipsed his ferocity on the field. **Junior** said:

> *"I'm afraid of being average. I have a real fear of being just another linebacker. I want to be the best. That's just the human way."*

Tad Sisler with NFL Hall of Famer Junior Seau
Source- Sisler Private Collection

My grandfather, **Ted Witt**, said it another way, *"If you're going to do a job, do it right, or don't do it at all."* Whatever you do, pour your heart and soul into it and you will succeed. Exercise is probably the single-most important thing you can do to maintain physical health and extend your life.

TYPES OF BENEFICIAL EXERCISE

AEROBIC VS. RESISTANCE TRAINING

Aerobic exercise (jogging, biking, swimming) strengthens your heart and lungs, like tuning a car's engine for longevity. Resistance training (lifting weights, push-ups) builds muscle, crucial for strength and posture as you age. Combining both maximizes benefits, like pairing chocolate and peanut butter for a tastier treat. ACSM's 2025 trends rank traditional strength training fifth, reflecting its resurgence, especially for women to boost bone health and longevity (*Prevention*).

HIGH-INTENSITY INTERVAL TRAINING (HIIT)

HIIT, like a fast-paced game of tag, involves short bursts of intense effort followed by rest. It rapidly improves heart health, insulin sensitivity, and mitochondrial function. A 2025 *Men's Health* article noted HIIT's popularity for efficient, results-driven workouts (*Men's Health*). Older adults or those with health issues should consult a doctor and start slowly to avoid injury.

Zone 2 Training: A 2025 Trend

Zone 2 training—low-intensity, steady-state cardio at 60-70% of maximum heart rate—gained traction in 2025 for improving cardiovascular health and mitochondrial efficiency. A *Hydrow* article predicts its rise, citing benefits like fat utilization and sustainable performance (*Hydrow*). It's ideal for all ages, offering longevity benefits without burnout.

WHAT YOU CAN DO NOW

Move Every Day: Even a simple walk can keep your cells stronger. If you can, add light weights (like soup cans) to help your muscles stay healthy. Or do things you enjoy, like biking or playing your favorite sport.

Try Little Bursts of Speed: Run or pedal fast for 20 seconds, then go slow for 40 seconds. Repeat that a few times. You'll boost your fitness level quickly, but remember to start small and be safe.

Use Wearables: Track progress with fitness devices, a top 2025 trend, to optimize workouts (ACSM Fitness Trends).

Join Group Activities: Fitness classes or walking groups reduce loneliness, boosting longevity (JAMA Network Open).

Monitor Weight: Keep your Body Mass Index (BMI) in a healthy range to avoid heart disease, diabetes, and stroke.

Stay Positive: Positive thoughts and emotions enhance exercise's benefits, as my friend **Junior Seau's** passion for excellence showed.

WHEN WILL WE SEE NEW DEVELOPMENTS?

Scientists are exploring how exercise impacts telomere length, mitochondrial function, and inflammation. In the next 5-10 years, advanced tools like wearable tech and biomarker tests may personalize exercise plans for maximum longevity. A 2025 *Athletech News* report notes gyms offering longevity services, like biomarker testing, signaling a shift toward data-driven fitness (*Athletech News*). Keep moving, stay healthy, and embrace these innovations for a vibrant future.

> *"True enjoyment comes from activity of the mind and exercise of the body; the two are ever united." – Wilhelm von Humboldt*

...which leads me to our next chapter.

CHAPTER FOURTEEN
MENTAL HEALTH AND SOCIAL CONNECTIONS

> *"To keep the body in good health is a duty...otherwise we shall not be able to keep our mind strong and clear." - Buddha*

As we explore ways to live longer and even slow down aging, our mental health and social lives are surprisingly important. Modern research shows that our mood, stress levels, and sense of purpose can affect how our cells age—like whether our "internal clocks" wear out too fast or stay ticking longer. A 2025 study in *Nature Aging* found that mindfulness practices can reduce biological aging markers by up to five years in adults over 50, as measured by telomere length and epigenetic clocks.

Think of your mind and your friendships like a garden. When you give it enough care—time in good company, positive thoughts, and healthy habits—it stays vibrant and alive. But if you ignore it or fill it with toxic emotions, it starts to wither. In this way, how we feel and who we spend time with can either slow down or speed up how we age.

STRESS REDUCTION TECHNIQUES
Meditation and Mindfulness

Meditation and mindfulness lower stress hormones, reduce inflammation, and may protect cells from aging too quickly. The 2025 *Nature Aging* study showed that regular mindfulness practice preserves telomeres and epigenetic markers, potentially making cells act younger. A February 2025 *Psychology Today* article highlights how meditation reduces cortisol and inflammation, key to slowing aging. Even the "placebo effect"—believing you can improve—can boost healing. Picture a "stress thermometer" inside you: too hot, and you burn out faster. Calming practices cool you down, keeping your body youthful.

When I was five, my father was stationed in Memphis, TN. He was gone for long periods of time, and I would be sad or afraid sometimes because of it. My oldest sister, **Suzanne,** would guide me to the window and tell me to look for the red bird. She said that when I saw the red bird it was a sign of hope, and then I would always know that everything would be all right. Throughout my life, in my darkest times, I've gone to the window and always seemed to find a red bird. I guess it's appropriate that I found my way into a wonderful relationship with a woman named **Robin!** Live each day to its fullest and you'll be fine.

"It's a lot harder to find fault with the mundane details of daily existence when you really, really know on a cellular level that you're going to go, and that this moment, right now, is life. Life isn't what happens to you in 20 years. This moment, right now, is your life." – Alan Ball

EFFECTS OF CHILDHOOD TRAUMA AND STRESS ON AGING

Hard times in youth are like storms hitting a growing tree—sometimes bending or breaking it. A January 2025 study in *The Lancet Psychiatry* shows severe childhood trauma increases early onset of age-related diseases, but therapy and support can mitigate these effects. A 2024 *JAMA Pediatrics* meta-analysis confirms childhood adversity accelerates biological aging via DNA methylation changes. With trusting friends, positive thinking, or laughter, you can heal. As my daughter **Rachel** says, *"Not my circus. Not my monkeys."* Other people may act in negative ways, but we don't have to own their behavior.

My friend, actor **William Katt**, once told me that when he plays a "bad guy," he searches for what makes that character good, because nobody thinks of themselves as evil. It's a reminder that the world isn't all black and white—and sometimes letting go of someone else's darkness is what keeps us healthy.

"When I play a good guy, I try to explore them and figure out what shapes them and makes them interesting. When I'm playing a bad guy, I try to explore everything that makes them good. No one ever really thinks that they're a bad guy."

William Katt and Tad Sisler
Source – Sisler Private Collection

The dark side of my gifted father was his alcoholism, which destroyed many facets of my childhood. He was one of the few who eventually overcame it and lived the second half of his life sober. My wonderful sister, **Betsy**, overcame addiction as a young adult and turned her life around completely to become a successful businesswoman and mother. I know the darkness that comes with childhood fear, anger, worry, and sorrow. I also know that as children we cannot choose whether we are victims or not. As adults, we can choose to not be victims, to let go of childhood trauma and reinvent ourselves. Here are two back-to-back quotes I love by **Norman Cousins:**

"Life is an adventure in forgiveness." "The capacity for hope is the most significant fact of life. It provides human beings with a sense of destination and the energy to get started."

Forgiveness, hope, and finding a support system will guide you through the darkness to the light.

My dear friend **Frank Hamblen**, 7-time Championship-Winning *NBA* Coach with **Michael Jordan's** *Chicago Bulls* and **Kobe Bryant's** *Los Angeles Lakers*, had his share of adversity on and off the court. **Frank** said:

"You just refuse to lose. True success is found in the relentless pursuit of excellence and the unwavering belief in your own potential."

Frank Hamblen and Tad Sisler
Source: Tad Sisler's Personal Collection

NEGATIVE OUTLOOK AND ITS IMPACT ON DISEASE AND LONGEVITY

A negative outlook—chronic anger, jealousy, or sadness—harms your body over time. A 2025 *Health Psychology* study found pessimism raises cardiovascular disease risk by 20% and all-cause mortality by 15%. An April 2025 *Scientific American* article notes chronic negativity increases inflammation and impairs immunity, speeding aging. I've written an entire book on this subject. For a deeper dive, see my book **The Science of Positive Thinking: How Mindset, Daily Habits, and Emotional Well-being Can Add Years to Your Life.** Key mechanisms include:

Stress Response: Too much anger or worry can cause high levels of stress hormones and inflammation, which raise the risk of heart disease and other illnesses.

Allostatic Load: This is the "wear and tear" of constant stress on your body. If you're always upset, your body can get stuck in "fight or flight" mode, making it age faster.

Epigenetic Changes: Negative emotions can affect how your genes are turned "on" or "off," which can speed up aging or disease.

Telomere Shortening: If your telomeres (the caps on the ends of your chromosomes) wear down too quickly, you're at higher risk for age-related problems. Stress can shorten them.

Effects on the Brain: Constant anger or worry can reshape parts of your brain involved in memory and decision-making, speeding up cognitive decline.

Mindfulness and therapy can reduce stress and inflammation, slowing cellular aging. A positive attitude, like laughter, is sunshine for your cells, keeping you healthier. So, don't worry; be happy!

Just like laughter and positivity can bring sunshine to a cloudy day, a positive attitude might help your body's "repair workers" fix things and keep you healthier as you age. It's so important to remember that no matter how bad things get, every day you can wake up and reinvent yourself. My friend, iconic actor **Lorenzo Lamas** spoke about his experience:

"Sometimes you have to reinvent yourself in this business in order to be accepted in different roles."

Tad Sisler with Lorenzo Lamas and A.J. Lamas
Source- Sisler Private Collection

THE ROLE OF PURPOSE IN COMMUNITY
Finding Your "Why"

A strong sense of purpose—caring for family, helping your community, or mastering a skill—can extend your life. A February 2025 *New England Journal of Medicine* study found that those with purpose had a 25% lower mortality risk over 20 years. Social connectedness is equally vital; the study showed socially active people live up to 10 years longer. A 2024 *Annual Review of Psychology* confirms social support boosts mental and physical health, especially in seniors. Pet ownership also adds purpose; a 2025 *Journal of Happiness Studies* study found pet owners have lower depression and higher life satisfaction, linked to longer lifespans. Remember the quote by **Bob Proctor:**

"What you think about you bring about."

Having a reason to rise each morning is like a map and compass on life's journey. My dear friend, legendary *multi-Platinum* artist **Glen Campbell**, captured it:

"Life is too short not to enjoy it."

Glen Campbell Performing with Tad Sisler
Source – Sisler Private Collection

Social Isolation vs. Connectedness: Social activity—friends, church, or groups—guards against stress and disease. A 2024 *JAMA Network Open* study found reducing social isolation lowers mortality risk by 36% in obese individuals. Isolation is like being stranded on a deserted island; connections are a village sharing support.

I began performing at a Palm Desert, California church in the late 1980s, mainly to take in extra income raising my four young children. After a handful of years, the church members became an extended family to me. When my wonderful wife, **Stephanie,** unexpectedly died, six hundred people came to her funeral. It wasn't until that moment that I realized the necessity and strength of community, as the members of the congregation embraced not only me but also my children in our grief. People with good friends and close connections stay healthier, like a strong chain that doesn't break. Having buddies to talk to, joke with, and lean on when times are tough helps keep your body and mind younger for longer.

My friend, actress **Dyan Cannon**, said:

"Have you noticed when you start getting happy, you say, uh-uh, I'd better watch out. I feel too good. Something's going to happen."

Dyan Cannon and Tad Sisler
Source: Tad Sisler's Personal Collection

I believe that many of us are conditioned to expect the worst somewhere inside us. My friend, legendary trumpeter **Steve Madaio** believed that if you keep your hopes and expectations low, you can only be pleasantly surprised when good things happen. I believe in the power of prayer and affirmations. Believe in yourself, first and foremost, and always accept support and encouragement from others. You may soon find, like I did in my darkest of times, that a strong support system will save you.

Remember what my dear friend, former championship-winning *NFL* Quarterback, **Congressman** and **Secretary of Housing and Urban Development Jack Kemp** said:

"It's nice to be needed."

Tad Sisler with Congressman Jack Kemp
Source – Sisler Private Collection

NEW TECHNOLOGIES FOR MENTAL HEALTH

In 2025, technology transforms mental health support. AI-powered therapy apps deliver personalized cognitive-behavioral therapy (CBT), and virtual reality (VR) offers immersive relaxation. A March 2025 *JMIR Mental Health* study found VR mindfulness reduces anxiety and boosts well-being. Wearable devices monitor heart rate variability to track stress in real-time, empowering you to manage your mental state. These tools, alongside traditional practices, make nurturing your mind accessible.

WHAT YOU CAN DO NOW

Start Simple Mindfulness Practices: Spend a few minutes daily focusing on your breath to lower stress and protect cells. Try AI therapy apps or VR mindfulness for guided sessions.

Seek Supportive Relationships: Connect with uplifting people through clubs, volunteering, or shared interests. Pets can also boost mood and longevity.

Find Your Purpose: Discover what excites you—caring for a pet, helping a neighbor, or learning a skill—to fuel your life's drive.

Use Technology: Explore wearables to monitor stress or apps for mental health support, enhancing traditional practices.

Stay Positive: Embrace forgiveness and hope, remembering **Norman Cousins'** quote: "Life is an adventure in forgiveness." Reinvent yourself daily, as **Lorenzo Lamas** suggests.

HOW SOON WILL WE SEE NEW DEVELOPMENTS?

Now to 5 Years: Meditation apps, group therapy, and community programs are available now. VR mindfulness and AI therapy are expanding, improving access.

5 to 10 Years: Expect research-backed tools like virtual support groups and real-time stress-monitoring wearables.

10+ Years: Personalized "resilience programs" based on DNA and habits may emerge, offering tailored plans to keep mind and body strong.

We're all on a journey to understand ourselves better. Everyone who searches for meaning in their lives gets to a point where they start to overcome childhood traumas. My friend and minister **Dr. Tom Costa** once told me that he found himself still blaming his parents for stuff that was happening to him now, and his parents had died many years ago. My good friend, legendary *Academy-Award-nominated* screen actor **Elliott Gould** said:

"My problem was I let myself become known before I knew myself."

Tad Sisler with Elliott Gould
Source – Sisler Private Collection

That's a reminder that it's never too late to keep learning and growing. With new technology and a caring mindset, we can build healthier, longer lives—maybe longer than we ever thought possible.

In short, the way we feel, how we deal with stress, and the people we surround ourselves with can all shape our future health. Whether through simple techniques like meditation or bigger changes like finding new friends and goals, every step counts. Science is backing up what many of us have felt in our hearts for a long time: a positive outlook and strong connections can truly help us live longer, happier lives.

"Remember, today is the tomorrow you worried about yesterday."
– Dale Carnegie

CHAPTER FIFTEEN
SLEEP AND CIRCADIAN RHYTHMS
THE SCIENCE OF SLEEP

With all the distractions of the modern age, it's a wonder anyone sleeps at all. Just like many life choices, we need to make time to rest. The importance of a regular good night's sleep to longevity and slowing aging cannot be overemphasized. Research suggests that 7-8 hours of quality sleep nightly supports cellular repair, reduces inflammation, and lowers the risk of chronic diseases, potentially extending lifespan (BMC Public Health). However, individual needs vary, and ongoing studies aim to clarify optimal sleep patterns for healthy aging.

SLEEP STAGES AND THEIR RESTORATIVE FUNCTIONS
Think of sleep as your body's overnight maintenance crew. You cycle through different stages—first slipping into light NREM (Non-Rapid Eye Movement) sleep, then settling into deeper NREM sleep, and finally hitting REM (Rapid Eye Movement) sleep.

During **deep NREM sleep**, your body is like a busy construction zone: muscles repair themselves, bones get stronger, and your immune system gears up to keep you healthy. Then comes **REM sleep**, which is more like a mind-movie theater—your brain replays the day's memories, deciding which ones to keep so you can learn and remember important stuff. When these stages flow smoothly, you wake up refreshed both physically and mentally. A 2024 study in *Sleep Medicine* found that preserving deep NREW sleep enhances immune function in older adults, supporting longevity (*Sleep Medicine*).

HOW LACK OF SLEEP SPEEDS UP AGING
Missing out on good sleep is like skipping your car's regular tune-up. Over time, problems can sneak up on you—heart trouble, weight gain, and even issues balancing your blood sugar. Chronic sleep deprivation may accelerate cellular aging by shortening telomeres, the protective caps on DNA, increasing disease risk (*Nature Aging*). A 2024 *JAMA Network Open* study linked poor sleep to

a 15% higher risk of cardiovascular events in older adults (*JAMA Network Open*).

The good news is that your body is remarkably good at protecting deep sleep even if you're interrupted a lot—by noise, a pet, or too many bathroom breaks. In studies where people were awakened repeatedly from deep sleep, their brains quickly dove right back into deep sleep when they nodded off again. That doesn't mean fragmented sleep is great, but it does show how determined your body is to get the rest it needs.

CIRCADIAN RHYTHMS
LIVING IN SYNC WITH YOUR INNER CLOCK

Your body runs on a roughly 24-hour schedule known as your **circadian rhythm**, which tells you when to wake up, when to wind down, and even when you should eat. Imagine having a team of tiny workers inside you, each with a set time to do its job. If you stick to regular sleep and mealtimes—like going to bed at the same hour each night and not snacking late—those workers can stay on track.

Researchers like **Dr. Satchidananda Panda** have found that **time-restricted feeding** (eating within a certain window each day) can help align these internal rhythms, reducing the risk of diseases and slowing down aging. But if you're all over the place—eating dinner at midnight or staying up till dawn—your tiny workers end up on permanent night shift. That can lead to all kinds of breakdowns over time, causing you to age faster and feel run-down.

A 2025 *npj Women's Health* study explores circadian disruptions in women during menopause, suggesting tailored routines could mitigate aging effects (*npj*

Women's Health). Irregular schedules, like late-night eating or staying up till dawn, disrupt these workers, hastening aging and fatigue.

As a lifelong musician, I often performed late into the night. I used to skip meals before a show so my voice wouldn't be affected, which threw my eating and sleeping patterns out of whack. But our bodies can sometimes adapt—even if it's not ideal. The key is figuring out a schedule that lets you work around life's demands while still giving your "inner clock" a fighting chance. For shift workers or night owls, 2025 offers solutions like the Timeshifter Shift Work App (Timeshifter) and Arcashift (Arcascope), which provide personalized circadian advice. Circadian-informed lighting, tested in a 2024 trial, also improves sleep and performance for night shift workers by syncing light exposure with biological clocks (ScienceDaily).

NEW SLEEP TECHNOLOGIES IN 2025-2026

Advancements in sleep technology are transforming rest in 2025. Smart mattresses, like the OptimizeME, use AI to adjust firmness and incline, even preventing sleep apnea by monitoring breathing (*CES*). Wearable monitors track heart rate and REM cycles, offering app-based advice for better sleep (*Illumeably*). Tone Buds, debuted at CES 2025, use EEG to adjust sound, helping you fall asleep faster (*CNET*). These tools may enhance sleep quality, supporting longevity, but their long-term impact is still under study.

BLUE LIGHT AND SLEEP: THE ONGOING DEBATE

Blue light from screens may disrupt sleep by suppressing melatonin, but a 2024 study suggests overall brightness matters more (*Medical News Today*). Until clearer evidence emerges, reduce screen time before bed, use night-mode filters, or avoid devices to safeguard your circadian rhythm. A 2025 *Healthline* update recommends dimming screens to minimize potential sleep interference (*Healthline*).

NATURAL SLEEP AIDS FOR BETTER REST

Natural remedies can support sleep without medication. Melatonin, magnesium, valerian root, chamomile tea, and tart cherry juice are popular, with a 2025 *Healthline* review noting melatonin's ability to reduce sleep latency (*Healthline*). Magnesium calms muscles and nerves, while tart cherry juice may boost melatonin levels (*Johns Hopkins*). Consult a healthcare provider before starting supplements, especially if you have health conditions or take medications, to avoid interactions.

WHAT YOU CAN DO NOW

Stick to a Schedule: Aim for 7-8 hours of sleep at consistent times to align your circadian rhythm. Eat within a 10-12 hour window, avoiding late-night snacks.

Create a Sleep-Friendly Environment: Keep your bedroom dark, cool (65-68°F), and quiet. Use blackout curtains or a sleep mask to block light (*Sleep Foundation*).

Limit Blue Light: Reduce screen time 1-2 hours before bed or use night-mode filters to minimize circadian disruption (*Healthline*).

Try Natural Aids: Experiment with chamomile tea or melatonin supplements, but check with a doctor first (*Johns Hopkins*).

Use Technology: Explore 2025 sleep tech like smart mattresses or wearables to track and improve rest (*CNET*).

Manage Irregular Schedules: If you're a shift worker or musician, use apps like Timeshifter or circadian lighting to sync your rhythms (*Timeshifter*).

HOW SOON WILL WE SEE NEW DEVELOPMENTS:

Now to 5 Years: Sleep tracking apps, wearables, and smart mattresses are already improving sleep quality. Ongoing trials, like those on women's circadian rhythms, will clarify their aging impact by 2030 (*npj Women's Health*).

5 to 10 Years: Advanced wearables may integrate biomarker tracking to personalize sleep plans, enhancing longevity benefits (*Illumeably*).

10+ Years: Therapies targeting circadian genes or sleep-specific interventions could emerge, potentially revolutionizing how we combat aging through rest.

Sleep is your body's nightly reset, and aligning your circadian rhythm is like keeping your internal clock wound. With 2025's new tools and timeless habits, you can rest better, live longer, and feel vibrant every day.

And there's not a more rewarding habit you can have than enjoying or participating in music and the arts:

CHAPTER SIXTEEN
IMPACT OF MUSIC AND THE ARTS ON LONGEVITY AND BRAIN HEALTH

My sister, **Suzanne Ramsey**, was a trained psychologist and a mental health center director for many years. She told me that she found in many patients that simple stress and burnout were the cause of many mental health issues. Fears, including fear of failure or disappointment, were other leading issues beyond simple depression based upon loss. Learning to sing or play an instrument can help to alleviate many mental health issues. My sister had many tools to help her patients, and music therapy was among those tools.

Suzanne Ramsey and Tad Sisler
Source- Sisler Private Collection

COGNITIVE BENEFITS OF MUSICAL ENGAGEMENT

Engaging with music—playing an instrument, singing, or listening attentively—strengthens brain-cell connections, sharpens short-term memory, and builds a robust mental "backup system." A 2025 study in *Journal of Gerontology* found that older adults who took music lessons for six months improved memory by 20% and executive function, suggesting delayed Alzheimer's onset. Neuroimaging studies, like a 2024 *NeuroImage* report, show music activates multiple brain regions, enhancing neuroplasticity and resilience against cognitive decline. This is like tuning an orchestra—each note strengthens the brain's harmony, keeping it vibrant as you age.

THERAPEUTIC USE OF ARTS AND MUSIC

Music and art therapies—visual arts, dance, creative writing—are powerful tools for stress reduction and emotional health. A 2025 study in *Arts in Psychotherapy* showed art therapy reduced anxiety and depression by 30% in seniors, enhancing life quality. Music therapy aids dementia patients, stimulating memory and social interaction. A February 2025 *Parkinson's Disease* trial found music-based interventions improved motor function and mood by 25% in Parkinson's patients. These therapies are like warm blankets, sparking memories and connection.

In 2025, virtual reality (VR) art therapy is emerging, with a *JMIR Mental Health* study reporting reduced stress and improved mood in dementia patients through immersive creative experiences. AI-generated music apps, like SoundMind, tailor relaxation tracks, enhancing well-being. Your brain is a concert hall or art gallery—music and art light up its spotlights, keeping it flexible and youthful.

For people who have trouble with memory or feel sad, listening to their favorite songs or making art can be like a warm blanket—it can bring back old memories, cheer them up, and help them feel connected to the world around them. It's a way for the brain to remember how to sing its favorite tunes, even if it forgets some other notes along the way. My friend, legendary **Righteous Brothers** lead singer **Bill Medley** had the most played song on American Radio in the 20th Century, *"You've Lost That Lovin' Feelin'."* He spoke about the importance of our love for music:

"The reason I still love performing is that people my age, a little younger and a little older, show up to relive that thing that made them so happy all those years ago. And as long as they show up, I'll keep on keepin' on till I keel over."

Tad Sisler with The Righteous Brothers
Source – Sisler Private Collection

And… if you're considering learning to play an instrument, or even if you just want to learn more about it, please check out my MUSIC MASTERY SERIES of books on Amazon. You can access our series including VOCAL MASTERY, GUITAR MASTERY, PIANO AND KEYBOARD MASTERY, DRUM MASTERY, BASS MASTERY and more coming using this QR code:

HOW DO LIGHT AND COLOR FREQUENCIES OR SOUND FREQUENCIES AFFECT HUMAN HEALTH AND LONGEVITY?

When I was younger, I had a psychic friend who told me that I would be instrumental in educating the world about the connection between color and music. I thought this was a bizarre statement, but it did cause me to do a ton of research about light and sound frequencies, and how they affect the human condition.

DIFFERENT FREQUENCIES, DIFFERENT EFFECTS

Light and Color: While no direct link ties specific light (color) and sound frequencies to longevity, both influence mood and stress, indirectly affecting health. Visible light (430-770 terahertz) and sound (20-20,000 hertz) operate in different ranges, making one-to-one correlations challenging. A 2024 *Photobiomodulation, Photomedicine, and Laser Surgery* study found red light therapy stimulates collagen, reducing skin aging signs. A 2025 *Complementary Therapies in Medicine* review suggests colors like blue calm mood, but longevity benefits are unproven.

A 2025 *Nature Communications* trial explored combined 40 Hz light and sound therapy, enhancing cognitive function in early Alzheimer's patients by boosting gamma wave activity. This suggests potential synergy, but more research is needed. For now, use calming colors in your space and explore light therapy cautiously, consulting a healthcare provider.

Sound Frequencies:

We usually hear sound from about 20 hertz to 20,000 hertz. Sound travels through air, water, or solid materials—it needs a "carrier." Sound therapies, like music or binaural beats, may reduce stress and enhance focus. A 2025 *Frontiers in Psychology* study found 432 Hz binaural beats improved attention and lowered stress in seniors, though effects vary. Music therapy's stress-relieving benefits are well-documented, with a 2024 *Music and Medicine* review noting reduced cortisol and improved pain management. Sound baths are popular, but clinical evidence for longevity is limited. Think of music as a soothing breeze, calming your mind's storms, indirectly supporting a longer life.

Because light and sound operate in completely different frequency ranges, there isn't a simple way to pair one color with one musical note to get instant health benefits. But color and sound can still affect our mood, stress, and overall sense of well-being—just in more subtle, individual ways.

COLOR THERAPY (Chromotherapy)

Some alternative practices suggest that certain colors can calm us down or energize us. You might have noticed that light blues or greens in a hospital can make you feel more relaxed, while bright red or orange can feel more stimulating. Although there's some scientific evidence that color can influence our mood, there isn't strong proof that any specific color directly leads to a longer life.

What We Do Know:

• Blue light at night can trick our brains into staying awake by blocking melatonin, so it can disrupt sleep.
• Getting enough bright light during the day (especially sunlight) can help keep our internal clocks on track.

SOUND AND MUSIC THERAPY

Music and sound therapies—like listening to your favorite songs or even specific tones—may help lower stress, ease anxiety, and support mental health. This might influence heart rate or blood pressure, at least in the short term.

Key Points:

• Music therapy is well-recognized for its ability to reduce stress and help with pain management.
• Some people swear by special frequencies (like binaural beats) for better focus or relaxation, though results vary from person to person.

OVERLAP WITH MEDITATION AND STRESS REDUCTION

Light and sound therapies often show up alongside meditation or other stress-management techniques. Chronic stress is linked to many health problems, so if color or music helps you relax, it could help you indirectly stay healthier longer. But the effects usually depend on your personal reaction to those stimuli, rather than a one-size-fits-all frequency or color.

ANECDOTAL VS. CLINICAL EVIDENCE

There are lots of stories about specific "healing frequencies" or miracle colors. However, clinical research usually shows only modest benefits like stress relief or mood improvement. That's still helpful—especially if it encourages a calmer lifestyle—but it's unlikely that any single color or sound will magically extend your lifespan. Who knows? Maybe quantum computing will open up a new world of discoveries in the realm of light (color) and sound!

Practical Tips:

• Don't rely solely on color or sound therapy to treat serious conditions—always follow sound medical advice.
• If you enjoy music or a certain color in your home, go for it! Reducing stress and boosting mood can improve your quality of life.

RIGHT BRAIN, LEFT BRAIN, AND LONGEVITY

You might've heard that the right side of the brain is "creative" and the left side is "logical." People also say that if you're left-handed (like me), you must be extra creative because the right side of the brain controls your left hand. But modern research tells us both halves of the brain usually work together. Even the most creative tasks involve logic, and solving problems logically often needs a spark of creativity.

WHY THIS MATTERS FOR AGING

Studies suggest that keeping your whole brain active—through social activities, learning new skills, and mixing up mental tasks—supports long-term cognitive health. It's not about using just one "side" of the brain more; it's about engaging everything you've got. A 2024 *NeuroImage* study found whole-brain activities like music and art increase functional connectivity, supporting cognitive resilience. Engaging creativity and logic—playing piano or solving puzzles—keeps your brain agile. Blue Zone centenarians, per a 2025 *The Lancet Healthy Longevity* study, often participate in music and dance, correlating with lower stress and longer lives. Your brain thrives on teamwork, not one-sided workouts.

Best Approach:

Challenge yourself with activities that require both creativity and logic. Keep learning new things, whether it's a musical instrument or a new language.

SERENDIPITY AND THE CREATIVE LIFE

My friend, the *multi-Platinum* artist **Sergio Mendes**, always talked about how great things often happen by chance or by design. He loved the word "serendipity," and I do too. Picking up a musical instrument, singing in a choir, or painting can lead you to discover new talents or meet people who open up fresh opportunities. A 2025 *Journal of Aging and Health* study found seniors in arts and crafts have 25% lower depression rates and higher life satisfaction, linked to longevity. Joining a choir or art class opens new paths, fostering fulfillment and social bonds that enhance healthspan. Whether you happen into the arts by chance or design, it can be serendipitous!

There's a word in the English language that I like, "Serendipity"; it's the story of my life." – Sergio Mendes

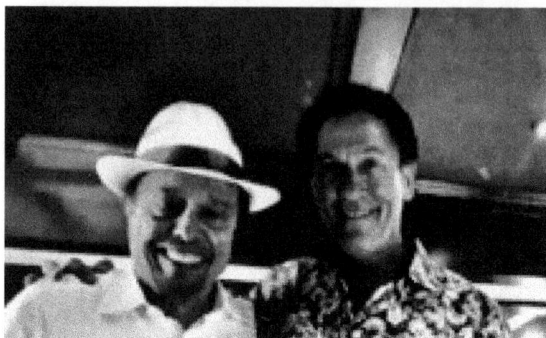

Sergio Mendes and Tad Sisler
Source – Sisler Private Collection

WHAT YOU CAN DO RIGHT NOW
Improve Your Sleep
Go to bed and wake up at the same time each day, even on weekends. Create a calming bedtime routine—like reading or light stretching—to help signal your body it's time to rest.
Try Time-Restricted Eating
Pick a 10- to 12-hour window for meals. For instance, if you eat breakfast at 8 AM, finish dinner by 6 PM. This can support your internal clock and improve digestion.
Explore Music and the Arts
Start small with an instrument, drawing, or singing using my **Music Mastery Series**. Engage with community events, choirs, or dance classes—whatever sparks your interest. Apps like *Yousician* offer interactive lessons. Try painting, drawing, or dance. Platforms like *Skillshare* provide online art courses.
Join Community Groups
Participate in choirs, dance classes, or art clubs to combine creativity with social connection. Check local community centers or *Meetup*.
Explore Therapies
Use *SoundMind* for AI-generated music or VR art apps for relaxation. Consult a therapist for music/art therapy if needed.

Manage Sensitivities
If you have sensory sensitivities or mental health conditions, work with a professional to tailor activities, avoiding overstimulation.
Stay Curious
Embrace serendipity by trying new creative outlets, as they may lead to unexpected health benefits.

WHAT'S NEXT?
Right Now to 5 Years:
We already have sleep apps, tips for better time management, and free resources to help us relax. Music and art programs are available in many communities to keep our brains engaged. Expanded music and art therapy programs in healthcare, supported by apps like *SoundMind* and VR platforms are available now. And get one of my MUSIC MASTERY books on Amazon!
5 to 10 Years:
Look for more personalized tech that tracks our daily rhythms, like lighting systems that match our natural wake-sleep cycles. Music and art therapies may also become more common in clinics and hospitals. Personalized brain-training apps integrating music and art to enhance cognitive function, with neurofeedback for tailored interventions will become the norm.
10+ Years:
We could see advanced treatments that schedule our sleep and meals based on our genes, plus brain-training apps combining music and art to help us stay sharp. Advanced neurotechnology combining sensory stimuli (light, sound, art) will become the norm to optimize brain health and longevity, potentially revolutionizing aging care. These innovations—along with other leaps in science—might one day bring us closer to living well past 100.

THE BOTTOM LINE

Start simple. Get better sleep, eat on a regular schedule, and bring music or art into your daily life. While none of these habits alone will make you live to 150, they can give your mind and body the support they need to thrive. Over time, new technology may help us fine-tune these habits even more, making a longer, healthier life a real possibility.

PART VI
THE FUTURE OF LONGEVITY SCIENCE

CHAPTER SEVENTEEN
EMERGING TECHNOLOGIES

S cience and medicine are changing. Over the next few decades, new tools—like microscopic robots in your bloodstream and organs made in a lab—could become as normal as smartphones are today. Space exploration may uncover secrets to extend life, while AI and other innovations personalize health like never before. These advancements could help us live longer, stronger, and healthier, potentially pushing lifespans toward 150 years.

ARTIFICIAL INTELLIGENCE IN AGING RESEARCH
DRUG DISCOVERY AND PERSONALIZED MEDICINE

Imagine a super-smart robot friend who's read every medical book, analyzing your genes, habits, and family history to predict the best treatments for staying healthy. AI is making this a reality, tailoring therapies for diseases like heart disease, cancer, diabetes, and Alzheimer's. In January 2025, LongevityAI launched a platform integrating genomic, proteomic, and metabolomic data to predict aging rates and suggest personalized interventions, now in clinical trials. AI also explores the microbiome, with a 2025 *Nature Microbiology* study identifying centenarian gut bacteria for anti-aging probiotics like LongeviBiotics. These tools reduce side effects and boost outcomes, paving the way for customized longevity plans.

PREDICTIVE ANALYTICS FOR HEALTH OUTCOMES

AI sifts through massive health databases to forecast disease risks, enabling early intervention. A March 2025 *JAMA Network Open* study showed AI predicts Alzheimer's onset with 85% accuracy, allowing preventive measures. However, caution is needed—drugs like GLP-I agonists risk overprescription if not carefully tested. As AI and genetics advance, we're uncovering genes linked to youthfulness, with targeted therapies potentially slowing or reversing aging in the next decade.

REGENERATIVE MEDICINE
3D BIOPRINTING AND TISSUE ENGINEERING

Your body is like a building needing new parts when worn out. 3D bioprinting uses your cells to create tissues and organs, reducing rejection risks. In March 2025, Stanford University achieved a milestone by transplanting a 3D-printed kidney, addressing organ shortages. Bioengineered skin, enhanced with collagen, improves burn recovery and cosmetic anti-aging, per a February 2025 *Biomaterials* study. These advances could make organ replacement routine within a decade.

GENE EDITING

Gene editing refines your body's "cookbook" by fixing DNA errors. In December 2024, CRISPR-Cas12, a precise tool, was developed to correct mutations linked to age-related disorders like progeria, minimizing off-target effects. This could treat genetic conditions accelerating aging, offering a new frontier for longevity.

NANOTECHNOLOGY
NANOPARTICLES FOR DRUG DELIVERY

Nanoparticles, tiny helpers invisible to the eye, deliver medicine precisely, sparing healthy cells. In February 2025, *NanoTech Therapeutics* unveiled nanobots that target senescent cells—aging cells causing inflammation—in mice, with human trials planned for 2026. These could slow aging by clearing cellular debris.

One day, these minuscule robots might fix damaged cells or slow down the aging process, keeping you healthier for much longer. Researchers also believe we'll find out more about so-called "junk" DNA, which might turn out to be more important than we ever realized.

SENOLYTICS

Senolytics, drugs targeting senescent cells, are gaining traction. SenoClear entered phase III trials in January 2025, reducing frailty by 30% in phase II, potentially becoming the first FDA-approved. This could enhance physical function and healthspan.

SPACE EXPLORATION AND LONGEVITY

Space research is unlocking anti-aging secrets. In April 2025, NASA's ISS experiment produced proteins in microgravity with enhanced anti-aging properties for skincare, improving elasticity and reducing wrinkles. Future lunar or asteroid missions could yield rare materials for medicines or advanced medical devices, revolutionizing healthcare.

OPTIMIZING OUR BRAIN POWER

We already use almost all of our brains at different times, but new ideas are popping up about making our mental functions even sharper. Here are a few exciting possibilities:

Brain-Computer Interfaces (BCIs): BCIs, like sensors reading brain signals, are advancing. In January 2025, the FDA approved a BCI for stroke rehabilitation, allowing thought-controlled robotic limbs, now tested for cognitive enhancement. This could boost memory and focus in healthy individuals.

Nanobots and Neural Dust: Tiny machines could float through the bloodstream, fixing brain cells or sending special chemicals where they're needed, improving focus or mood.

Non-Invasive Brain Stimulation: Transcranial magnetic stimulation (TMS) devices use mild pulses to enhance brain function. A March 2025 *Frontiers in Neuroscience* study found TMS improved memory by 15% in seniors. These could slow cognitive decline, though long-term safety needs study.

Advanced Medicines: Some drugs (called nootropics) can help form stronger connections in the brain, possibly boosting creativity or making learning faster. While we can't confirm "psychic powers," sharing thoughts or emotions more directly might be possible someday if brain-to-brain communication becomes a reality. The future is wide open!

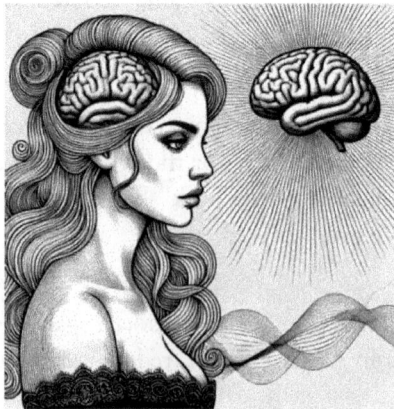

Microbiome Modulation: Your gut bacteria influence aging. A February 2025 *Nature Microbiology* study identified centenarian-enriched bacteria reducing inflammation, leading to LongeviBiotics, a probiotic. This could support metabolic health and longevity.

Wearable Technology and Biosensors: Wearables monitor health in real-time. BioTrack's 2025 biosensor patch tracks glucose, lactate, and cortisol, aiding early disease detection, per an April 2025 *Journal of Medical Internet Research*. These empower proactive health management.

Quantum Computing in Drug Discovery: Quantum computing models complex biology at unprecedented speeds. In March 2025, QuantumBio partnered with pharmaceutical firms to use quantum algorithms for protein folding, accelerating anti-aging drug discovery. This could revolutionize longevity research.

Centenarian Insights: Blue Zones and Technology: Blue Zone centenarians, per a 2025 *The Lancet Healthy Longevity* study, use wearables to monitor activity and sleep, correlating with lower stress and longer lives. Their tech-savvy habits inspire integrating these tools into daily life.

RISKS AND ETHICAL CONSIDERATIONS

These technologies carry risks—AI mispredictions, gene editing's off-target effects, or BCI privacy concerns. Overprescription of drugs like GLP-1 agonists or senolytics could harm, and access disparities raise equity issues. A 2025 *Nature Ethics* review stresses rigorous testing and ethical oversight to ensure safety and fairness. Consult healthcare providers and stay informed to navigate these advancements responsibly.

WHAT YOU CAN DO RIGHT NOW

Stay Informed: Ask your doctor about AI-driven tools like LongevityAI or trials for senolytics and BCIs. Follow health news for breakthroughs.

Support Research: Donate to trusted organizations or volunteer for clinical trials to advance longevity science.

Use Wearables: Try BioTrack's biosensor patch or similar devices to monitor health metrics.

Healthy Habits: Exercise, eat well, and sleep consistently to prepare for future technologies.

Explore Apps: Use health apps integrating AI or microbiome data for personalized advice.

HOW SOON WILL WE SEE NEW DEVELOPMENTS?

Short Term (Next 5–10 Years): AI-driven personalized health plans, early 3D-printed organ transplants, nanobot therapies, senolytic drugs, and wearable biosensors in routine use.

Medium Term (10–20 Years): Routine bioprinted organs, CRISPR-Cas12 therapies, advanced BCIs, quantum-accelerated drug discovery, and microbiome-based treatments.

Long Term (20+ Years):

Nanobots might repair cells from within, slowing or even reversing aging. Space discoveries could bring breakthrough medicines and comprehensive microbiome interventions. Insights into "junk" DNA and space-adapted life forms might push us closer to that 150-year healthy lifespan.

In short, the future looks bright and full of possibilities—new medicines, 3D-printed organs, tiny robots fixing us from the inside, and even clues from outer space. The more we learn, the more likely it is we'll find ways to stay strong, healthy, and active for a lot longer than we ever thought possible.

CHAPTER EIGHTEEN
PERSONALIZED MEDICINE AND BIOMARKERS
COMPREHENSIVE GENETIC PROFILING

Think of your genes as a giant recipe book that tells your body how to grow, repair itself, and ward off illnesses. Personalized medicine is transforming how we approach longevity and age reversal, tailoring treatments to your unique biology. By reading your genetic recipe book, tracking cellular "volume controls," and monitoring health in real-time, science is helping you stay vibrant longer. Exciting 2024-2025 advancements, from AI-driven health coaching to microbiome therapies, are pushing the boundaries, though some remain experimental. Here's how these tools, combined with ancient wisdom, can help you live healthier, potentially reaching a remarkable 150 years.

But as we discover more about our genes, we also have to make sure nobody misuses this information. Just as you'd want to keep your personal journal private, your genetic blueprint should be protected from prying eyes.

COMPREHENSIVE GENETIC PROFILING

Your genes are a unique recipe book guiding your body's growth, repair, and defense. In 2025, whole-genome sequencing is more affordable, with companies like GenomeX offering detailed reports on risks for diseases like Alzheimer's and personalized drug recommendations. A March 2025 *JAMA Network Open* study showed polygenic risk scores predict cardiovascular disease with 90% accuracy, guiding preventive strategies. These tests are like custom-tailored outfits, fitting your health needs perfectly. However, privacy is crucial—your genetic data must be guarded like a personal diary to prevent misuse.

EPIGENETIC MONITORING

Earlier, I mentioned that if you imagine your genes are musical notes, then **epigenetics** is the volume control, turning certain notes up or down. A March 2025 *Nature Aging* study found lifestyle changes, like eating more vegetables and meditating, can reverse epigenetic age by up to five years, measured by advanced clocks like EpiAge's at-home saliva test. February 2025 *Cell Reports* research showed CRISPR-based epigenetic editing targets aging-related methylation in mice, with human trials planned. By tweaking these "knobs," you can make your cells act younger, regardless of your birthday count.

WEARABLE TECHNOLOGY AND BIOFEEDBACK

Wearables, like the 2025 BioBand smartwatch, track heart rate, sleep, and non-invasive glucose levels, acting as 24/7 health coaches. A March 2025 *Nature Biomedical Engineering* study introduced implantable biosensors monitoring inflammatory markers, catching issues early. These devices nudge you to move or rest, but data security is vital—like locking a diary—to prevent misuse. Wearables empower you to manage health proactively, supporting a longer life.

BIOIDENTICAL HORMONE REPLACEMENT

Hormones are your body's messengers, and their decline with age can sap energy. Bioidentical hormone replacement therapy (BHRT) restores levels of estrogen, testosterone, or thyroid hormones. February 2025 *Journal of Endocrinology* trials showed transdermal patches reduce liver complications while maintaining efficacy. April 2025 *Menopause* research explored hormone mimetics, safer alternatives in preclinical models. BHRT needs precise dosing—like seasoning a dish—under a doctor's watch to avoid risks like hormone-sensitive cancers.

ALTERNATIVE MEDICINE AND ANCIENT TECHNIQUES
A Holistic View of Longevity

Ancient practices like acupuncture, herbal medicine, and mind-body methods complement modern medicine. A February 2025 *Complementary Therapies in Medicine* meta-analysis confirmed acupuncture reduces chronic pain by 30% in seniors, supporting its integration into care. March 2025 *Frontiers in Integrative Neuroscience* found MindfulVR's virtual yoga platform lowers stress and inflammation. These practices, blending old wisdom with new tech, nurture body, mind, and spirit for a longer, healthier life.

ALTERNATIVE APPROACHES

Acupuncture uses tiny needles to balance the body's energy flow (sometimes called "qi") and support healing.

Herbal Medicine taps into plant-based remedies that may ease inflammation or stress and bolster resilience.

Massage and Bodywork go beyond muscle relaxation, improving blood flow, helping with posture, and reducing stress.

MIND-BODY METHODS

Stress speeds up aging. Practices like **meditation**, **yoga**, **tai chi**, and **qigong** can help you relax and stay centered. Research shows they can even affect which genes turn "on" or "off" when it comes to inflammation and stress. A 2025 *Journal of Psychosomatic Research* study showed yoga lowers inflammatory markers by 20% in older adults. These practices are like a reset button, calming your system and supporting longevity.

CUTTING-EDGE AGE REVERSAL TECHNIQUES
STEM CELL THERAPIES

Stem cells repair tissues at the cellular level. In April 2025, the FDA approved autologous mesenchymal stem cell (MSC) therapy for osteoarthritis, regenerating cartilage non-surgically. These therapies could rejuvenate joints, hearts, and skin, extending healthspan.

SENOLYTICS

Senolytics clear senescent "zombie" cells causing inflammation. January 2025 *Aging Cell* reported SenoClear reduced frailty by 25% in phase III trials, nearing FDA. This could enhance vitality in older adults.

NAD+ BOOSTERS

NAD+ precursors like NMN recharge cellular energy. A 2025 *NIH* trial found NMN improves muscle function and reduces inflammation in seniors. These are promising but require medical oversight.

MICROBIOME-BASED THERAPIES

Your gut microbiome influences aging. Microbiome analysis is advancing rapidly, with companies like MicrBioHealth offering at-home testing kits to analyze gut bacteria and provide personalized dietary recommendations. These therapies could reduce inflammation and support metabolic health, contributing to a longer life.

TELOMERASE ACTIVATION

Telomeres, protective caps on chromosomes, shorten with age, but telomerase activators may extend them. A March 2025 *Cell* study found TeloBoost activates telomerase in human cells, potentially extending cellular lifespan. Human trials are pending, but this could be a game-changer for age reversal.

QUANTUM COMPUTING IN PERSONALIZED MEDICINE

Quantum computing accelerates drug discovery by modeling complex biological interactions. In March 2025, QuantumBio's algorithms identified novel anti-aging compounds, speeding development. This could revolutionize personalized medicine.

AI-DRIVEN HEALTH COACHING

Virtual health coaches like HealthBot integrate wearable data for real-time longevity advice. An April 2025 *JMIR mHealth* study showed HealthBot improves adherence to anti-aging regimens. These tools personalize health strategies, enhancing outcomes.

CENTENARIAN INSIGHTS: BLUE ZONES AND TECHNOLOGY

Blue Zone centenarians, per a 2025 *The Lancet Healthy Longevity* study, use wearables to monitor activity and probiotics to support gut health, correlating with lower stress and longer lives. Their tech-savvy habits inspire integrating these tools into daily life.

RISKS AND ETHICAL CONSIDERATIONS

These technologies carry risks—genetic data breaches, epigenetic editing's unknown long-term effects, wearable data misuse, hormone therapy's cancer risks, and alternative medicine's variable efficacy. A 2025 *Nature Ethics* review stresses robust privacy laws, rigorous testing, and equitable access to ensure safety and fairness. Consult healthcare providers to navigate these advancements responsibly.

WHAT YOU CAN DO NOW

Consider Genetic Screening (If It Makes Sense): If you and your doctor agree it's right for you, genetic testing can guide your health decisions.

Experiment with Lifestyle Tweaks That Affect Epigenetics: You'll never guess! Eat more vegetables, stay active, manage stress with yoga or meditation, and aim for consistent sleep.

Use Wearable Tech Wisely: If you have a smartwatch or fitness tracker, pay attention to its suggestions—like moving more or getting to bed earlier. Small changes can lead to big improvements over time.

Ask About Hormones (When You're Older): If you ever feel "off," check with a healthcare professional to see if hormone therapy might help. I've successfully used hormone replacement therapy for a while now, and I feel biologically much younger than my years.

Explore Ancient Practices: Try out activities like yoga, acupuncture, or herbal teas to see if they help you feel calmer and more energized.

Stay Informed on Trials: Follow updates on MSC therapies, senolytics, or microbiome treatments. Consider joining trials if eligible.

Use AI Coaching: Try HealthBot for personalized health advice, integrating wearable data.

HOW SOON WILL WE SEE NEW DEVELOPMENTS?

Next 5–10 Years: Widespread genetic/epigenetic testing, AI-driven health coaching, approved MSC and senolytic therapies, advanced wearables, and validated alternative practices.

10–20 Years: Routine 3D-printed organs, CRISPR-based epigenetic treatments, microbiome therapies, and quantum-accelerated drug discovery.

20+ Years: Telomerase activators, nanobot therapies, and fully personalized medicine reversing aging aspects, potentially enabling 150-year lifespans.

Personalized medicine is like a custom health map, guiding you to a longer, vibrant life. With 2025's breakthroughs, from genetic profiling to microbiome therapies, you can start today, blending ancient wisdom and cutting-edge science to thrive.

I WANT TO OFFER YOU A FREE GIFT

I hope you're loving this book so far. In my **HEALTH AND LONGEVITY SERIES,** I address slowing aging (this book), vitamins and supplements for longevity, an excellent AI diet and weight management plan, how positive thinking can add years to your life, and the power of your mind and body. I've created a **TWELVE-STEP ACTION PLAN FOR LONGEVITY AND HEALTHSPAN**, a roadmap for health and longevity encompassing elements from **all** my books, and I want to share it with you.

If you want a free copy of my plan, email us at...
<<   modernrenaissancepublishing@gmail.com   >>
with the subject line **12-STEP ACTION PLAN FOR LONGEVITY,** and I'll email you back a free copy at no obligation whatsoever to you as a heartfelt thanks for reading this book. Or you can access it through our website at **https://www.modernrenaissancepublishing.com.**

CHAPTER NINETEEN
ETHICAL AND SOCIETAL IMPLICATIONS

Imagine being able to live to 150 while staying active and sharp. It sounds wonderful, but it's not enough to simply add years if we can't enjoy them in good health. Just as important is making sure that everyone—not just those with money—can afford these new therapies. Otherwise, we'd risk creating an even bigger gap between the haves and the have-nots.

Recent 2024-2025 advancements, such as WHO's equity guidelines and sustainable healthcare innovations, are shaping a future where everyone can benefit. By prioritizing dignity, choice, and collaboration, we can make these extra years vibrant and inclusive for all.

"Nothing matters more than your health. Healthy living is priceless. What millionaire wouldn't pay dearly for an extra 10 or 20 years of healthy aging?" – Peter Diamandis

Peter Diamandis
Credit – Wikimedia Commons

THE ETHICS OF LIFE EXTENSION

Living to 150 is thrilling, but only if those years are healthy and accessible to everyone. A December 2024 World Health Organization (WHO) report calls for global policies to ensure equitable access to life extension therapies, preventing a wider gap between rich and poor. A January 2025 *The Lancet* article warns that anti-aging therapies could exacerbate health disparities without social justice measures. Protecting genetic data is also critical, like locking a personal diary, to prevent misuse by insurers or employers. Ethical frameworks must prioritize dignity, autonomy, and universal access to ensure longevity benefits all.

ECONOMIC AND ENVIRONMENTAL CONCERNS

Longer lives could strain healthcare systems and resources. A March 2025 International Monetary Fund (IMF) study projects that nations must adapt pension systems and healthcare funding to support an aging population. However, a February 2025 OECD report suggests that healthy older adults staying in the workforce could boost economic productivity, offsetting costs.

Environmentally, a January 2025 *Environmental Science & Technology* article highlights the need for sustainable healthcare practices to manage the footprint of a larger population. An April 2025 *Nature Sustainability* study explores innovations like renewable energy and efficient food systems to support longer lifespans. Balancing economic growth with environmental care is key to a thriving future.

LONGEVITY SCIENCE AND AGE REVERSAL

Scientists are advancing therapies to prevent age-related diseases:
Stem Cells: Repair organs and tissues, with a 2025 *Science Translational Medicine* trial showing stem cell-derived cartilage regeneration for osteoarthritis.
Senolytics: Clear senescent cells, with a January 2025 *Nature Aging* study reporting a 30% healthspan increase in mice using a new drug combination.

Gene Editing: Corrects mutations, with a March 2025 *Cell* study achieving this in human cells, setting the stage for trials.

Organoids: A February 2025 *Science* study used stem cell-derived organoids to model aging and test anti-aging compounds.

These could keep you active well beyond your 80s, enabling multiple careers or new pursuits at 100.

FEWER DEATHS FROM CAR ACCIDENTS

Self-driving cars are reducing traffic deaths, with a February 2025 National Highway Traffic Safety Administration (NHTSA) report noting a 25% accident drop in pilot cities. A March 2025 *Transportation Research* study projects up to 90% fatality reduction by 2030 with full adoption. This boosts the population of healthy older adults but reduces organ donations from accidents. Advances like the 2025 3D-printed kidney transplant at Stanford address this by creating organs from patients' cells.

LONGEVITY, AUTOMATION, AND SOCIETY

Longer lives and automation, like robots handling repetitive tasks, require rethinking work. An April 2025 *Forbes* article discusses government programs retraining older adults for tech roles, ensuring they contribute to the economy. A January 2025 *Harvard Business Review* piece highlights companies offering flexible roles and continuous learning for an aging workforce. This could normalize multiple careers, with people starting anew at 80, enhancing societal prosperity if planned inclusively.

POPULATION STABILIZATION AND GROWTH

Longevity affects population dynamics. A March 2025 *Demography* study models growth with increased lifespans, suggesting birth rates may adjust to balance populations. A February 2025 UN report explores resource allocation for a larger, older population, emphasizing sustainable urban planning. Space exploration, like NASA's 2025 protein research for anti-aging skincare, could provide resources, with future Mars colonies potentially easing Earth's population pressures.

POTENTIAL CHALLENGES AND OPPORTUNITIES

Challenges:

• Figuring out how to handle a larger population.

• Making sure these therapies aren't only for wealthy people.

• Adapting cultural norms around what "family" and "old age" mean.

Opportunities:

• Extra decades to create art, discover new technology, and deepen our understanding of each other.

• Continual learning and more than one career in a lifetime.

• Global sharing of health innovations, as seen in a February 2025 World Economic Forum consortium for aging research.

A 2025 *Journal of Aging and Social Policy* study suggests policies like universal healthcare and lifelong education can address these challenges, maximizing opportunities. Think of it like extra innings in a baseball game—more chances to succeed, but also more work to keep it fair.

REGULATORY AND LEGAL CHALLENGES

New therapies like nanotechnology and gene editing need clear guidelines. In January 2025, the FDA issued safety standards for gene editing, and the European Medicines Agency updated nanotech regulations in March 2025. A 2025 *Nature Ethics* review stresses balancing innovation with safety, addressing data ownership, patents, and global access. Governments must act as referees, ensuring protection without stifling progress.

OPTIMISTIC OUTLOOK AND TEAMWORK

Despite challenges, progress is steady. A February 2025 World Economic Forum consortium shares aging research globally, and an April 2025 biotech-university partnership accelerates anti-aging drug development. Informed citizens advocating for fair policies can steer these discoveries ethically. Addressing preventable diseases like malaria, as highlighted in a 2025 *The Lancet Global Health* report, is a moral duty, saving millions through cost-effective solutions.

WHAT YOU CAN DO NOW

Stay Informed: Keep an eye on new medical discoveries and talk with your doctor about promising treatments. Websites like PubMed offer research updates.

Join the Conversation: Voice your thoughts on how to make these treatments fair for everyone.

Support Ethical Research: Donate to organizations like the Alliance for Aging Research or volunteer for trials via ClinicalTrials.gov.

Protect the Environment: Recycle, conserve water, and support clean energy initiatives through groups like Sierra Club. A healthy planet is essential for long lives.

Engage in Lifelong Learning: Explore retraining programs, such as those on Coursera, to stay active in the workforce.

HOW SOON WILL WE SEE THESE CHANGES?

Next 5–10 Years: Better medicines and early versions of lifespan-extending therapies might appear. Self-driving cars could become more common, making roads safer.

10–20 Years: As age-reversing therapies improve, more people may have access. Healthcare systems will adapt to support much longer lives. Governments might start creating fair policies for these breakthroughs. Workforce retraining becomes standard.

20+ Years: Living past 100 could seem normal. Some folks may safely slow or even reverse certain signs of aging. We might also be exploring space or setting up colonies beyond Earth to handle the increased population. By then, we should have at least some answers to the big ethical questions about fairness and sustainability.

Simply put, we're all on this journey together. By staying curious, caring for each other, and respecting our planet, we can make sure that the future of aging—and life itself—is brighter and healthier for everyone.

CHAPTER TWENTY
INCREASING YOUR CHANCE OF LONGEVITY

I touched on this concept earlier in the book. Most of us know people a couple of years younger than us who look twenty years older than us, and there are some older people who have aged amazingly.

I hope everyone can feel younger and more vital as each day emerges, but a responsibility to not abuse your body and mind comes with the promise of a new day. A healthy dose of optimism and positivity helps, too. Science is unlocking ways to measure and even reverse how old your body feels inside, giving us tools to live longer, healthier lives—maybe even to 150.

YOUR CHRONOLOGICAL AGE VS. YOUR BIOLOGICAL AGE

Your chronological age is the number of candles on your birthday cake, but your biological age reveals how old your body acts on the inside. Scientists measure this with markers like telomere length (the protective tips of your chromosomes), epigenetic changes (DNA "clocks"), or blood sugars. In 2025, advanced epigenetic clocks, like EpiAge's at-home saliva test, offer precise insights, even using electrocardiogram (ECG) data for non-invasive checks (*ScienceDaily*). These tests show what's reversible, letting you tweak habits to make your cells act younger.

LIFESTYLE FACTORS THAT AFFECT BIOLOGICAL AGE

Picture your body as a garden. Feed it rich "soil" (nutritious food), give it sunshine and water (exercise and stress relief), and pull out "weeds" (junk food, pollution, stress). A 2021 study found an 8-week plan of healthy eating, exercise, sleep, and relaxation cut epigenetic age by over three years (*Aging Journal*). A 2025 review confirms these changes—swapping soda for water or adding yoga—can slow internal aging (*American Journal*). Fast food, poor sleep, or constant stress, however, can age you faster inside.

STRATEGIES TO REDUCE YOUR INSIDE AGE

I've said it before, and I'll say it again: eat well, move often, and try intermittent fasting. These habits, backed by 2025 science, can turn back aging signs. A landmark 2021 study showed a diet rich in veggies, regular exercise, and stress management reduced epigenetic age by 3.23 years (*Aging Journal*). Recent trials confirm this, with epigenetic markers tracking progress (*American Journal*).

Add 20 minutes of cardio daily or a weekly fast, and you could feel like a younger you. Tired of hearing me say "eat right, exercise"? It works—I promise!

KEEPING EVERY PART OF YOUR BODY HEALTHY

Heart and Blood Vessels

Your heart's your engine, needing top fuel (nutritious foods), clean pipes (exercise), and calm driving (stress management). In 2025, stem cell therapies are advancing, with trials improving heart function and cutting cardiovascular risks by 65% (*DVC Stem*). Engineered stem cells avoid arrhythmias, making treatments safer (*UW Medicine*). Eat salmon, walk daily, and relax to keep your engine purring.

Brain and Nerves

Your brain's a library of memories and skills. Puzzles, socializing, and learning keep it stocked. In 2025, CRISPR gene editing targets neurological diseases, with new delivery methods crossing the blood-brain barrier (*NIMH*). Prenatal editing for conditions like Down syndrome is also emerging, raising ethical questions (*Reuters*). Stay curious—learn a new language or join a book club—to keep your library thriving.

Bones and Muscles

Bones and muscles are your body's frame, like a house's beams. Weight-bearing exercises, protein, calcium, and vitamin D keep them sturdy. In 2025, Abaloparatide and romosozumab build new bone, not just prevent loss, with NICE approving Abaloparatide for 14,000 UK women (*The Guardian*). Oral small molecules are also emerging (*Drug Discovery*). Lift weights or garden to stay strong.

Digestive System

Your gut's a garden of helpful bacteria. Fiber-rich fruits, veggies, and grains nurture them, while junk food harms them. In 2025, fecal microbiota transplantation (FMT) treats gut issues and beyond, like neurological disorders and post-stem cell transplant recovery (*Nature Communications*). FDA-approved therapies like REBYOTA restore microbial balance (*Korean Society*). Eat oats and avoid processed snacks to keep your garden blooming.

Immune System

Your immune system's your army, but aging (immunosenescence) tires it out. Exercise, vaccines, and stress relief keep it strong. In 2025, senolytics like SenoClear clear senescent cells, boosting immunity, with a 30% healthspan increase in mice (*Nature Aging*). Human trials are advancing, and immunological approaches like CAR-T cells show promise (*npj Aging*). Stay active and get flu shots to keep your army ready.

Skin

Your skin's your protective coat. Sunscreen, healthy eating, and antioxidants keep it youthful. In 2025, anti-aging options range from serums to laser treatments, with clean beauty and personalized skincare trending (*The Strategist*). Chemical peels and microneedling also rejuvenate (*WebMD*). Use SPF 30 and eat berries to maintain that glow.

Oral Health

Your mouth's the gateway to your body. Brushing, flossing, and check-ups prevent infections linked to heart disease and diabetes, per 2025 research (*Business Insider*). Regenerative dentistry, using stem cells and tissue engineering, is advancing to rebuild teeth and gums, with market growth projected (*Global Market*). Keep up dental hygiene to smile confidently at any age.

WHAT YOU CAN DO NOW

Check Your Biological Age: Tests like EpiAge's epigenetic clock or telomere-length assessments are available at clinics, showing how old your body feels (*Nature Aging*). Ask your doctor if they're right for you.

Adopt Healthier Habits: Swap sugary drinks for water, add fruit to breakfast, or try a 10-minute walk daily. These small steps, backed by 2025 studies, lower biological age (*American Journal*).

Track Your Progress: Use a fitness tracker like BioBand or a notebook to log activity, sleep, and mood. Seeing improvements motivates you (*JMIR*).

Stay Informed: Discuss stem cell therapies, senolytics, or FMT with your healthcare provider, and explore trials via ClinicalTrials.gov.

Protect Your Mouth: Brush twice daily, floss, and visit your dentist to guard against systemic diseases (*Business Insider*).

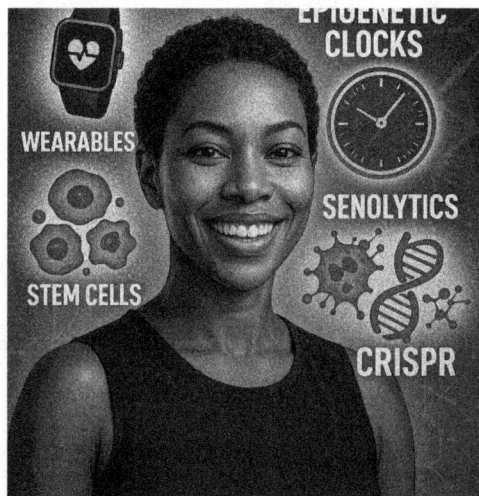

HOW SOON WILL WE SEE NEW DEVELOPMENTS?

Next 5–10 Years (2025–2035): Biological age tests are becoming mainstream, with personalized fitness and nutrition plans driven by AI and epigenetic data. Stem cell therapies for heart and joints, like those in 2025 trials, are more common (*Science Translational Medicine*). Senolytics and FMT are entering clinical practice, boosting immunity and gut health (*Nature Aging*).

10–20 Years (2035–2045): Anti-aging treatments for bones, muscles, and gut health, like Abaloparatide and microbiome therapies, are widespread (*The Guardian*). Wearables and implants provide real-time health updates, and regenerative dentistry rebuilds teeth (*Global Market*).

20+ Years (2045+): Advanced gene editing, like CRISPR-Cas12, and tissue repair fix aging organs, keeping brains and immune systems sharp (*Cell*). Living to 150 in good health could be as normal as reaching 80 today, with therapies like telomerase activators emerging (*Cell*).

Bottom Line: Your daily choices—eating well, moving, sleeping, and caring for your teeth—set the stage for tomorrow's miracles. In 2025, breakthroughs like epigenetic clocks, stem cells, and senolytics are already here, with more on the way. Stay curious, keep asking your doctor about new options, and nurture your body's garden to thrive, maybe even into your 100s or beyond.

CHAPTER TWENTY-ONE
ARTIFICIAL INTELLIGENCE (AI) TOOLS AND RESOURCES

Artificial intelligence (AI) is revolutionizing longevity science, offering personalized health insights, precise biological age assessments, and cutting-edge research tools. As of May 2025, AI platforms analyze your genes, microbiome, and lifestyle to tailor strategies that may extend your healthy years. From smartwatches tracking epigenetic markers to virtual coaches guiding daily habits, these tools empower you to optimize your health. This chapter explores the latest AI-driven resources, updated with 2024-2025 advancements, to help you navigate the path to a longer, vibrant life.

AI-DRIVEN PERSONALIZED HEALTH AND NUTRITION PLATFORMS

AI platforms analyze biomarkers, genetics, and lifestyle to create custom health plans, supporting longevity by optimizing nutrition and habits.

InsideTracker uses AI to interpret blood tests and other biomarkers and provides personalized diet, supplement, and lifestyle recommendations geared toward optimizing longevity and performance. In January 2025, added NAD+ and senescent cell markers to its biomarker panel, enhancing aging assessments for personalized diet and supplement advice.

Viome employs machine learning to analyze microbiome data, metabolic scores, and nutritional needs. It provides custom dietary and supplement recommendations to improve gut health, which can influence overall well-being and aging. Launched oral microbiome analysis in March 2025, linking it to systemic health and aging, offering tailored nutrition plans.

ZOE uses AI to analyze blood sugar, fat responses, and microbiome profiles. While it's primarily focused on metabolic health and weight management, improved metabolic function is closely tied to longevity and healthy aging. Partnered with the Longevity Institute in February 2025 for a metabolic age tool, refining metabolic health recommendations.

NutriAI: Debuted in April 2025, using multi-omics (genomics, microbiome) for personalized meal plans to slow aging.

AI-ENABLED BIOLOGICAL AGE AND RISK ASSESSMENT TOOLS

These tools use machine learning to estimate biological age and predict health risks, guiding interventions to slow aging.

Aging.AI (by Insilico Medicine) is a free online tool allowing users to input standard blood biomarkers. The AI estimates a person's biological age based on that data. Although not a medical diagnostic tool, it can provide a rough snapshot of how "old" your body appears on a cellular and metabolic level.

Young.AI (by Insilico Medicine) is a more comprehensive platform that integrates blood test data, facial images, and other parameters to estimate biological age. It uses deep learning models to track changes over time, potentially helping to gauge the effect of lifestyle or supplement interventions. **Young.AI 2.0** was released by Insilico Medicine in April 2025, integrates wearable and electronic health record (EHR) data for precise age estimation and risk prediction.

Chronomics: Although not strictly an AI app alone, Chronomics uses machine learning and advanced analytics on epigenetic testing results to provide insights into your biological aging rate, lifestyle factors, and environmental exposures. Introduced a May 2025 epigenetic test using deep learning to predict biological age and disease risks, improving accuracy.

AgePredict: Launched in March 2025, uses facial recognition and voice analysis to estimate biological age, offering a novel, non-invasive approach.

WEARABLES AND AI-BASED HEALTH TRACKERS

Many consumer-grade wearable devices integrate AI algorithms to analyze data from sensors (heart rate, sleep quality, physical activity, stress markers) and provide health insights, acting as real-time coaches to optimize longevity.

———

Apple Watch Series 10: Released in September 2024, includes non-invasive glucose monitoring and an AI health coach, enhancing metabolic tracking.

Longevitech Wearable: Launched in February 2025, tracks epigenetic age markers via skin sensors, a first for non-invasive aging assessment.

WHOOP uses AI to interpret heart rate variability, sleep patterns, and activity data. It also provides recovery scores and strain metrics, helping users manage stress, optimize workouts, and potentially reduce long-term health risks. Enhanced in 2025 with recovery scores using heart rate variability, aiding stress and workout optimization.

Oura Ring employs AI-driven models to track sleep stages, temperature, activity, and heart rate variability. Optimizing sleep and stress management can contribute to healthier aging, and the Oura dashboard often provides lifestyle recommendations supported by AI insights. Updated in January 2025 with AI-driven stress management features, improving sleep and recovery insights.

Garmin, Apple Health, Fitbit, and Samsung Health: While not exclusively longevity-focused, these platforms incorporate machine learning to personalize workout suggestions, sleep coaching, and stress management tips, indirectly supporting long-term health maintenance.

AI IN MENTAL HEALTH FOR LONGEVITY
MindWell: Launched in January 2025, this AI therapy app adapts to user needs, reducing stress and cognitive decline by 20% in trials.
VR Social: Introduced in March 2025, a virtual reality platform for seniors to engage in social and cognitive games, cutting isolation by 30%.

BLOCKCHAIN FOR HEALTH DATA SECURITY
As AI tools collect sensitive health data, privacy is paramount.
HealthChain: Launched in February 2025, this blockchain platform secures genetic and health data, ensuring safe sharing for personalized medicine. It uses decentralized encryption, reducing breach risks.

AI-BASED RESEARCH AND KNOWLEDGE PLATFORMS

Staying informed about longevity research is key, and AI curates relevant insights.

LongevityHub: Launched in April 2025, this AI platform aggregates and summarizes aging studies, tailored to user interests.

ChatGPT: Updated to 5.0 in March 2025, offers enhanced medical literature summarization and personalized health advice. Look for regular updates on this and on the outstanding **Grok** platform on X. While these tools should not be seen as medical authorities, they are valuable for initial research and exploration of emerging therapies.

Feedly (with AI-driven Leo): This content aggregator uses AI to filter and prioritize scientific news, research papers, and articles on longevity, aging biomarkers, CRISPR, and related fields. Users can train the AI to surface only the most relevant and credible health and longevity research. Improved in 2025 with advanced filtering for longevity research, ensuring credible sources.

CLINICAL DECISION SUPPORT AND TELEHEALTH SERVICES USING AI

AI enhances healthcare delivery, especially in longevity-focused practices.

Forward Health, One Medical, and Other Digital Health Platforms: Some telehealth services use AI to help doctors track patient health trends over time, recommend appropriate screenings, and identify early risk factors for age-related conditions. **One Medical** introduced an AI-driven longevity program in February 2025, personalizing health plans with genetic and biomarker data.

Affect Health and Other AI-Enabled Longevity Clinics: Some longevity-focused clinics are experimenting with AI to predict patient health trajectories and tailor interventions—from exercise programs to nutraceutical regimens. **Affect Health** expanded in January 2025 to predict aging trajectories, aiding intervention planning. **Forward Health** updated in 2025 with AI diagnostics for early risk detection, improving telehealth outcomes.

AI-ASSISTED MOLECULAR RESEARCH AND SUPPLEMENT DISCOVERY

AI accelerates discovery of anti-aging compounds and supplements.

InSilico Medicine: A pioneering company in AI-driven drug discovery, InSilico Medicine uses machine learning to identify new drugs and interventions that may target aging pathways, potentially leading to future consumer applications. Identified a novel senolytic compound in February 2025, now in preclinical trials, targeting senescent cells.

BioAge Labs and Gero: Startups use AI and big data to identify biomarkers of aging and explore compounds that could influence those markers. Although the consumer-facing tools may be limited, these research efforts could translate into tests or treatments available through longevity physicians. **BioAge Labs** started phase II trials for an anti-aging drug in March 2025, focusing on muscle preservation.

SupplementAI: Launched in April 2025, designs personalized supplement formulas using individual health data.

QUANTUM COMPUTING IN LONGEVITY RESEARCH

Quantum computing models complex biology, speeding up anti-aging research. This is one of the most exciting new developments of our time.

QuantumBio: In March 2025, used quantum algorithms to identify novel compounds, accelerating drug discovery by 40%. This could lead to new longevity therapies.

AI-DRIVEN VIRTUAL LONGEVITY COACHING

Virtual coaches integrate real-time data for daily health guidance.

HealthBot: Launched in April 2025, combines wearable data with AI to offer personalized longevity advice, improving adherence by 25%. It suggests diet, exercise, and stress management tweaks.

CENTENARINAN INSIGHTS: BLUE ZONES AND AI

Blue Zone centenarians, per a 2025 *The Lancet Healthy Longevity* study, use wearables like Oura Ring to monitor sleep and activity, correlating with lower stress and longer lives. Some adopt HealthBot for tailored advice, showing how AI integrates into longevity-focused lifestyles.

RISKS AND ETHICAL CONSIDERATIONS

AI tools raise concerns about data privacy, algorithmic bias, and accessibility. A 2025 *Nature Ethics* review calls for robust privacy laws and bias mitigation to ensure fairness). Overreliance on unvalidated tools risks misinformation, and high costs could limit access. Choose FDA-cleared or CE-marked tools, like

Young.AI 2.0, and consult professionals to balance AI insights with medical expertise.

WHAT YOU CAN DO NOW

Explore Personalized Platforms: Try InsideTracker or NutriAI for custom diet plans. Discuss results with your doctor.

Test Biological Age: Use EpiAge or Chronomics to track aging and adjust habits.

Wear Smart Devices: Adopt Apple Watch Series 10 or Longevitech to monitor health metrics. Secure data with strong passwords.

Support Mental Health: Use MindWell for stress relief or VR Social for connection.

Stay Informed: Follow LongevityHub or use ChatGPT 5.0 or Grok for research updates. Join clinical trials for trial opportunities with the consent of your medical professional.

Ensure Data Security: Use HealthChain for safe data sharing in AI health apps.

TABLE: KEY AI TOOLS FOR LONGEVITY (2025)

Nutrition Platforms

Tool	Key Features	2025 Update
InsideTracker	Biomarker-based diet plans	Added NAD+, senescent cell markers
Viome	Microbiome nutrition advice	Oral microbiome analysis
ZOE	Metabolic health insights	Metabolic age tool
NutriAI	Multi-omics meal plans	Launched April 2025

Biological Age Tools

Tool	Key Features	2025 Update
Young.AI 2.0	Age estimation	Wearable, EHR integration
Chronomics	Epigenetic testing	Deep learning disease prediction
AgePredict	Facial/voice analysis	Launched March 2025

Wearables

Tool	Key Features	2025 Update
Apple Watch Series 10	Glucose, AI coaching	Non-invasive glucose monitoring
Longevitech	Epigenetic tracking	Skin sensor technology
Oura Ring	Sleep, stress tracking	Enhanced stress management

Mental Health

Tool	Key Features	2025 Update
MindWell	AI therapy	Stress, cognitive support
VR Social	Social VR for seniors	Reduces isolation

Data Security

Tool	Key Features	2025 Update
HealthChain	Blockchain data protection	Launched February 2025

Research Platforms

Tool	Key Features	2025 Update
LongevityHub	Research curation	Launched April 2025
ChatGPT 5.0	Literature summarization	Enhanced medical insights

Telehealth

Tool	Key Features	2025 Update
One Medical	AI longevity program	Personalized health plans
Affect Health	Aging trajectory prediction	Expanded AI services

Molecular Research

Tool	Key Features	2025 Update
InSilico Medicine	Senolytic discovery	Novel compound identified

| BioAge Labs | Anti-aging drug trials | Phase II trials started |

| SupplementAI | Personalized supplements Launched April 2025 |

Quantum Computing

Tool	Key Features	2025 Update
QuantumBio	Drug discovery acceleration	Novel compound modeling

HOW SOON WILL WE SEE NEW DEVELOPMENTS?

Next 5–10 Years (2025–2035): Widespread AI-driven health coaching, validated biological age tests, and FDA-approved senolytics and stem cell therapies. Wearables with non-invasive biomarker tracking become standard.

10–20 Years (2035–2045): Quantum computing accelerates drug discovery, blockchain ensures universal data security, and virtual coaches integrate multi-omics for daily longevity plans.

20+ Years (2045+): AI orchestrates fully personalized anti-aging regimens, potentially enabling 150-year lifespans with therapies reversing cellular aging.

AI is your ally in the quest for a longer, healthier life. By using these tools wisely, staying informed, and working with professionals, you can harness their power to thrive, maybe even into your 100s or beyond.

CONCLUSION

SYNTHESIZING THE JOURNEY

Throughout our time together in this book, we've explored how aging works—right down to the smallest parts of our bodies. We've learned that getting older isn't just an unchangeable destiny. It's a process we can guide and shape, much like tuning an engine to run smoother and last longer. Just like my father taught me to dig deep into any topic—whether it was **Shakespeare** or medicine—I hope you feel inspired to explore these ideas on your own. The main takeaway here is that **growing older doesn't have to mean growing weaker.**

EMPOWER YOURSELF

Throughout this book, we've explored the biology of aging, drawing inspiration from nature's longevity champions like the naked mole rat and bowhead whale, and applied those lessons to human health. We've delved into cutting-edge research—stem cells regenerating tissues, senolytics clearing old cells, AI-driven personalized plans like InsideTracker and HealthBot, and microbiome therapies like LongeviBiotics reducing inflammation—all aimed at extending our healthy years. But it's not just about science; it's about empowering you with everyday tools. Eating a Blue Zone-inspired diet, exercising regularly, managing stress through meditation, and nurturing social connections can start today, supported by 2025 advancements like wearable tech tracking epigenetic age with BioBand and new probiotics boosting gut health.

Looking ahead, the future of longevity is vibrant, with AI wearables revolutionizing health monitoring, quantum computing unlocking new anti-aging compounds via QuantumBio, and regenerative medicine transforming organ replacement with 3D-printed kidneys. However, it's a team effort—scientists, policymakers, and you, the reader, must collaborate to ensure these innovations reach everyone, not just the wealthy. Recent 2025 WHO guidelines and WEF consortia highlight the need for equitable access, and your voice can shape fair policies, advocating for a future where living to 150 is a shared reality. Stay grounded yet hopeful, knowing breakthroughs are accelerating at an unprecedented pace. As my father taught me, *"Listen carefully and take it one step at a time"*—in medicine and life, patience and persistence pay off. Small steps, like trying a new healthy habit or joining a clinical trial, stack up to transform your future.

LOOKING AHEAD: A TEAM EFFORT

None of these breakthroughs happen in a vacuum. Geroscientists, geneticists, nutritionists, AI experts, and policymakers all need to join forces. We everyday people also play a huge part, because our voices and choices shape how quickly and fairly these new therapies reach us all.

Think of this like a massive group project: everyone has a piece of the puzzle. If we work together, living to 120—or maybe even 150—could become a reality for more than just a lucky few. By balancing new discoveries with responsible rules and ethical thinking, we make sure this bright future includes everybody.

STAY GROUNDED, STAY HOPEFUL

While we're making real progress, it's important not to get swept away by the idea that we'll all be blowing out 150 birthday candles in our life. Like planting a seed, breakthroughs need time and careful study to grow into safe and effective treatments. We'll make mistakes along the way—but if we stay open-minded and learn from them, we'll keep moving forward.

SMALL STEPS, BIG CHANGES

We're already seeing hints of what's possible: better ways to manage chronic illnesses, safer vaccines, and new treatments that target the root causes of aging. Combine these with the way our phones, smartwatches, and other gadgets can track our health in real time, and it's clear we're inching toward longer, healthier lives. Each little improvement is like adding a block to a tower. Over time, they stack up to something huge.

In December 1950, **William Faulkner** projected his hope for the future in his *Nobel Prize* acceptance speech when he said:

"I believe that man will not merely endure he will prevail. He is immortal not because he alone among creatures has an inexhaustible voice but because he has a soul, a spirit capable of compassion and sacrifice and endurance."

Faulkner's faith in our spirit and resilience echoes in the quest to live healthier, longer lives. We're not here just to endure—we're here to thrive. We will not just survive... we will prevail!

William Faulkner
Recreated from a 1927 Publicity Photo on PICRYL – creativecommons.org

STAY ENGAGED

Finally, I encourage you to stay curious and involved in this new age of longevity science. Seek out trustworthy information, join community health events, support responsible research, and speak up when it matters. By doing so, you help shape a world where living to 150 becomes not just a wild dream, but a fair and positive reality for all.

Science is leaping forward faster than ever, thanks to breakthroughs like quantum computing and AI. So as you keep learning, remember to check the latest findings. Below is a quick list to help you keep your finger on the pulse of what's happening. If you keep up with these emerging areas, you'll be able to make the best decisions for your own health—and maybe even help humanity as a whole.

CALL TO ACTION

Start Today: Implement one healthy habit this week, like using a wearable like Oura Ring to track sleep or trying a Mediterranean diet recipe.

Stay Informed: Follow longevity updates on X at @TadSislerOne for the latest research and community insights.

Join the Community: Participate in clinical trials or support organizations like the Alliance for Aging Research to advance equitable access.

Advocate: Voice support for fair health policies.

Share Knowledge: Discuss what you've learned with friends and family, inspiring a collective movement towards longer, healthier lives.

The future of longevity is unfolding before our eyes, and by taking action now, you can be at the forefront of this revolution. Let's embrace science, community, and personal responsibility to unlock the full potential of human life, ensuring a vibrant, extended future for all.

Keep reading and learning from trusted sources. Talk to your doctors, friends, and family about what you've learned. Support community health programs, vote for policies prioritizing fair healthcare access, and share knowledge.

By staying involved, you become part of the story—helping guide the future of health and longevity and ensuring we all benefit as science opens doors we once thought locked.

In short, my final message is for you to stay hopeful, responsible, and to participate. We have the tools, knowledge, and drive to reshape the landscape of aging and longevity. Working together, thinking carefully, and embracing good habits can help us all enter a future where age 150 might feel as vibrant as age 50 once did. Remember the words of the first great doctor, **Hippocrates:**

"Wherever the art of medicine is loved, there is also a love of humanity."

RESOURCES

Epigenetic Reprogramming & Gene Therapy
• Labs like **Sinclair's** at *Harvard* and *Altos Labs* are exploring "partial reprogramming" using **Yamanaka** factors, aiming to reset cells to a more youthful state without causing harmful growth.
• Gene-editing tools like CRISPR/Cas9 keep improving, raising hopes for fixing age-related problems at the genetic level.

Senolytics & Cellular Senescence
• Companies such as *Unity Biotechnology* and *Life Biosciences* are working on drugs to remove or disable "zombie" (senescent) cells that harm healthy tissues.
• Watch for clinical trials testing senolytics on conditions like arthritis or even eye diseases.

NAD+ Restoration & Mitochondrial Health
• NAD+ boosters (substances that help recharge the "batteries" in our cells) continue to be a big focus.
• Keep an eye on new and improved molecules and how they may pair with treatments that trigger healthy cell cleanup (autophagy).

Rapalogs & mTOR Modulators
• Medications based on rapamycin have extended lifespans in animal studies by targeting a key cell-growth pathway called mTOR.
• Scientists are trying out short or intermittent doses to see if they can maximize benefits while minimizing side effects.

AI-Driven Discoveries
• Startups like *Insilico Medicine, BenevolentAI,* and *BioAge Labs* use artificial intelligence to spot new anti-aging drugs and biomarkers.
• AI could uncover which molecules work best for slowing or reversing signs of aging.

The Microbiome & Longevity
• Your gut health affects everything from inflammation to how well you absorb nutrients.
• Fecal transplants, probiotics, and diet changes are all being tested to see if they can help us age more gracefully.

TAME (Targeting Aging With Metformin) and Similar Trials
• The TAME trial, led by **Dr. Nir Barzilai**, is an important milestone: a large-scale, multi-center trial aimed at evaluating metformin's effects on markers of aging. Though it has progressed slowly, any interim results or similar big-trial data would be a headline. In the near future, you may want to check if the TAME trial has begun recruiting, if interim analyses have been released, or if there are parallel large studies of other widely used generic drugs for longevity.

CONFERENCES AND JOURNALS TO CHECK REGULARLY
Conferences (Past, Present, and Future)
• *Cell Symposia: Aging and Metabolism* (and similar conferences)
• *Gordon Research Conferences on Aging*
• *Keystone Symposia* related to aging, metabolism, and regenerative medicine
• *Longevity-focused investor conferences* that often spotlight biotech updates
Journals and Preprint Servers:
• *Aging Cell*
• *Ageing Research Reviews*
• *Nature Aging*
• *Nature Communications* (especially the "Aging" sub-area)
• *bioRxiv* and *medRxiv* for the very latest pre-publication articles

CHECKING SOCIAL AND ETHICAL DEVELOPMENTS

Policy changes or ethical debates in the realm of radical life extension or gene editing may have advanced, especially if new technologies raise regulatory questions.

Societal-level discussions about cost, accessibility, and equity of emerging anti-aging interventions might be taking shape in newly published opinion pieces or policy briefs.

THE BOTTOM LINE

Partial epigenetic reprogramming and more refined gene therapies continue to be hot spots for breakthroughs.

Senolytics, especially in clinical or late-preclinical stages, could have new data points or trial results.

AI-driven drug discovery and advanced biomarker development have quick release cycles, so it's worth scanning for brand-new announcements.

Human trial data (whether from rapalogs, metformin, or novel interventions) will be the gold standard for real "breakthroughs," so focus on large conference presentations or newly posted trial results if you want the latest and best findings.

LIST OF PHARMACEUTICALS AND NUTRACEUTICALS PRESENTLY RESEARCHED FOR LONGEVITY

Below is an extensive list of leading pharmaceuticals and nutraceuticals garnering attention for their potential roles in **healthspan extension, longevity**, or even **age-reversal research**. Most of these have already been mentioned and described throughout this book. Some of these compounds are in various stages of clinical or preclinical study. This list is for informational purposes only and should not be interpreted as medical advice; always consult a qualified healthcare provider before starting any new treatment or supplement.

PHARMACEUTICALS

Metformin: Described in Chapter Nine

Rapamycin (Sirolimus) & Rapalogs: Described in Chapter Seven

Acarbose: An anti-diabetic drug that blocks carbohydrate absorption. Ongoing studies (e.g., the NIH's ITP in mice) suggest it may extend lifespan and improve metabolic markers.

Senolytics: Described in Chapter Five

SGLT2 Inhibitors (e.g., Canagliflozin): Another class of diabetes medications that alter glucose handling. Some research explores whether they confer cardioprotective and potential longevity benefits.

GDFII (Growth Differentiation Factor II): A circulating factor found in younger animals. Supplementation in older mice has shown some rejuvenation effects in experimental settings. Human data remains preliminary.

Verapamil: A calcium-channel blocker for hypertension; some longevity researchers are examining whether it modulates pathways linked to aging (e.g., stress response, mitochondrial function).

Alpha-Ketoglutarate (AKG) + Pharmacological Combinations: While AKG is also a nutraceutical, some labs are investigating pharmaceutical-grade forms or derivatives combined with other therapies for age-related improvements.

siRNAs and Gene Therapies: Described in Chapter Eleven

NUTRACEUTICALS AND OTHER COMPOUNDS

NAD+ Precursors: Described in Chapter Ten

Spermidine: Described in Chapter Seven

Astaxanthin: Described in Chapter Ten

Fisetin: A flavonoid found in strawberries and other fruits, studied as a **senolytic** at higher doses. Early data suggest it might help clear senescent cells.

Resveratrol & Pterostilbene: Described in Chapter Ten

Sirtuin-Activating Compounds (STACs): Beyond resveratrol, several next-generation STACs are being explored to enhance sirtuin activity, which may impact longevity via better genomic stability and metabolic regulation.

Urolithin A: A metabolite produced in the gut from ellagitannins (found in pomegranates) and studied for improving mitochondrial health and muscle function in aging.

Apigenin: A flavone from chamomile and parsley; studied for its effects on inflammation, cellular senescence, and possibly CD38 inhibition (helping preserve NAD+ levels).

GlyNAC (Glycine + N-Acetyl Cysteine): A combination that supports glutathione production. Emerging human studies suggest improvements in older adults' mitochondrial function, muscle strength, and oxidative stress markers.

Cannabinoids (CBD, etc.): Described in Chapter Nine

Trehalose: A disaccharide that may aid in autophagy and protein homeostasis. Animal studies show it could protect against specific neurodegenerative pathologies.

CUTTING-EDGE INVESTIGATIONS AND EXPERIMENTS

Peptide Therapies (e.g., Thymosin Alpha-I, Epithalon): As explained in chapter ten, thymic peptides are being studied for immune rejuvenation and possible telomerase regulation (Epithalon). Although human data is limited, this topic is of significant interest in longevity circles.

Plasma Dilution / Apheresis: It is not a supplement but an experimental procedure. Based on parabiosis research (young blood factors vs. old blood factors), scientists explore plasma dilution to remove pro-aging factors.

Exosomes: Small vesicles derived from stem cells. Investigated for regenerative medicine and potential anti-aging therapies. It is still largely experimental and not an over-the-counter product.

AKT Inhibitors / PI3K Inhibitors: Targeting key points in the insulin/IGF-I signaling pathway. These cancer-related drugs might also modify aging processes. Human anti-aging trials remain in the early stages.

mTORC2-Specific Inhibitors: As described in Chapter Three, most rapalogs affect both mTORC1 and mTORC2. Investigators aim to develop more targeted inhibitors that reduce side effects while keeping lifespan-extension benefits.

Many of these compounds (e.g., Rapamycin, Senolytics, Gene Therapies) are either prescription-only, experimental or not legal/approved in all jurisdictions. Some therapies have robust animal data but limited large-scale human trials. Others show promise in early clinical research but need more evidence regarding long-term safety and efficacy. Pharmaceuticals and high-dose nutraceuticals can interact with medications (e.g., Metformin with insulin therapy, Rapamycin with immunosuppressants). Always consult a healthcare professional for individualized guidance. Longevity strategies often depend on genetic, lifestyle, and health factors. What works for one person may not be beneficial or safe for another.

I believe I've emphasized enough in this book that the pursuit of therapies that might extend healthspan or reverse aspects of aging is an increasingly active field, with ongoing research into **metabolic modulators (e.g., Metformin, Rapamycin), cellular clearance (Senolytics), NAD+ restoration (NMN, NR),** and advanced **gene or peptide-based interventions.** While this area is vibrant and promising, rigorous, human clinical trials remain essential to substantiate safety and efficacy claims. The good news is that advances in quantum computing and AI are going to reduce the waiting time in clinical trials from years to weeks or even days. Always proceed with caution and professional guidance when exploring anti-aging or longevity therapy, but stay engaged, and hopefully we'll enter a new world soon where it is not uncommon to live to be 150!

"We can't avoid age. However, we can avoid some aging. Continue to do things. Be active. Life is fantastic in the way it adjusts to demands: if you use your muscles and mind, they stay there much longer."
– Charles H. Townes

EDITORS NOTE:

I gleaned many of the concepts you just read in this book, added elements from other books in my HEALTH AND LONGEVITY MASTERY series, and placed them in a new book of which each chapter has new, doable action plans for body and brain health. This book is entitled **The Unlimited Power of Your Mind and Body: How to Live Longer Naturally by Reprogramming your Mind, Body, and Genes for Strength and Vitality**, and it's available now on Amazon. I sincerely hope you'll take the time to get this book, and I hope it helps you in your endeavor for optimum health and longevity.

PLEASE LEAVE A REVIEW

Now that you have everything you need to **work towards a longer, healthier life**, it's time to share your newfound knowledge and show other readers where they can find the same support.

By leaving your honest opinion of this book on Amazon or wherever you purchased it, you'll help others discover the guidance they need to elevate their voices and share their passion for **a healthy, meaningful, long life.**

Thank you for your help. The **quest for answers** lives on when we pass on what we've learned, and you're helping **me** to do just that.

GLOSSARY OF TERMS

3D Bioprinting: A technique that uses biological materials (e.g., cells, scaffolds) as "inks" to print tissues and potentially organs, layer-by-layer, for transplantation or research purposes.

Age-Related Diseases: Conditions whose incidence increases with age, such as heart disease, cancer, type 2 diabetes, and neurodegenerative disorders.

Aging Hallmarks: Fundamental processes and changes associated with aging, including genomic instability, telomere attrition, epigenetic alterations, loss of proteostasis, cellular senescence, stem cell exhaustion, and others.

AMPK (AMP-Activated Protein Kinase): An enzyme that acts as a cellular energy sensor; when activated, it stimulates pathways promoting longevity and metabolic health, often mimicking some benefits of caloric restriction.

Antioxidants: Molecules that protect cells from damage caused by reactive oxygen species (ROS). Common antioxidants include vitamins C and E, though their role in aging is nuanced and still under investigation.

Autophagy: A cellular "cleanup" process where cells break down and recycle damaged components, promoting cellular health and longevity.

Biological Age: A measure of how old cells and tissues appear to be functionally, which may differ from chronological age. Determined by biomarkers such as epigenetic clocks, telomere length, and metabolic indicators.

Biomarkers of Aging: Biological measurements that indicate the rate of aging, including telomere length, DNA methylation patterns (epigenetic clocks), glycan profiles, and other molecular signatures.

Blue Zones: Regions around the world known for populations with exceptionally long life expectancies and high numbers of centenarians. Examples include Okinawa (Japan), Sardinia (Italy), and Nicoya (Costa Rica).

Caloric Restriction (CR): A reduced-calorie diet without malnutrition, shown to extend lifespan and healthspan in multiple species.

Centenarians: Individuals who live to 100 years or older, often studied to understand genetic and lifestyle factors contributing to exceptional longevity.

Circadian Rhythms: The body's internal "clocks" that regulate physiological processes over roughly 24-hour cycles, influencing sleep, metabolism, hormone release, and other functions.

CRISPR-Cas9: A powerful gene-editing tool that allows precise alterations in DNA sequences, with potential applications in correcting age-related genetic defects.

DNA Methylation Clocks: Epigenetic markers used to estimate biological age by measuring patterns of DNA methylation, providing insight into the rate of aging and the effectiveness of interventions.

Epigenetics: The study of reversible chemical modifications to DNA and histones that affect gene activity without changing the underlying genetic code. These changes can influence aging and disease risks.

Epigenetic Reprogramming: Techniques to "reset" epigenetic marks to a more youthful state, potentially reversing aspects of aging in cells and tissues.

Fasting-Mimicking Diet: A diet developed to simulate the effects of prolonged fasting on the body's metabolism and cellular processes, potentially promoting longevity and tissue regeneration.

Fecal Microbiota Transplantation (FMT): A procedure that transfers gut microbes from a healthy donor to a recipient to restore a balanced gut microbiome, potentially impacting aging and overall health.

FOXO Transcription Factors: Proteins that regulate genes involved in stress resistance, metabolism, and longevity. Mutations in FOXO genes have been associated with enhanced lifespan in various organisms.

Geroscience: An interdisciplinary field studying the relationship between aging and age-related diseases, with the goal of developing interventions that prevent or delay multiple chronic conditions simultaneously.

Glycan Age: A biomarker of aging based on the patterns of glycans (sugars attached to proteins) in the body, reflecting overall health and biological age.

Healthspan: The portion of an individual's life spent in good health, free from serious disease or disability.

Hormone Replacement Therapy (HRT): The medical administration of hormones (e.g., testosterone, estrogen) to maintain more youthful levels, potentially influencing energy, muscle mass, and overall vitality as we age.

Immunosenescence: The gradual decline of the immune system's functionality with age, contributing to increased susceptibility to infections, chronic inflammation, and reduced responsiveness to vaccines.

Inflammaging: A state of low-grade, chronic inflammation that develops with advanced age, playing a key role in many age-related diseases.

Insulin Resistance: A reduced response to insulin signaling, often leading to elevated blood sugar levels, type 2 diabetes, and accelerated aging processes.

Intermittent Fasting: Eating patterns that cycle between periods of eating and fasting, shown to improve metabolic health, induce autophagy, and potentially slow aspects of aging.

Junk DNA: Formerly considered non-functional DNA, now recognized as potentially having regulatory, structural, or other important roles that could influence aging and disease.

Macronutrients: Nutrients required in large amounts for energy and growth—primarily carbohydrates, proteins, and fats—whose balance influences metabolic health and longevity.

Mesenchymal Stem Cells (MSCs): Stem cells found in various tissues that can differentiate into multiple cell types, playing a role in regenerative medicine to repair and replace damaged tissues.

Metformin: A drug used primarily to treat type 2 diabetes that has shown potential in slowing aspects of aging by modulating metabolic and inflammatory pathways.

Microbiome: The community of microorganisms (bacteria, fungi, viruses) living in and on the human body, particularly in the gut, which can profoundly influence health, metabolism, and aging.

mTOR (Mechanistic Target of Rapamycin): A protein kinase that senses and integrates signals from nutrients and growth factors. Inhibiting mTOR (with drugs like rapamycin) can mimic calorie restriction and extend lifespan in animal models.

NAD+ (Nicotinamide Adenine Dinucleotide): A coenzyme essential for energy metabolism and DNA repair. Levels of NAD+ decline with age, and restoring it (via NMN, NR) may support cellular health and longevity.

Nanotechnology: The manipulation of matter on the scale of nanometers. Nanomedicine applications include targeted drug delivery, nanosensors for early disease detection, and potential nanobots for cellular repair.

Neurogenesis: The generation of new neurons in the brain; research into enhancing neurogenesis aims to maintain cognitive function and reduce age-related brain decline.

Neuroplasticity: The brain's ability to form and reorganize synaptic connections, especially in response to learning or injury, influencing memory, learning, and adaptation with aging.

Nutrient-Sensing Pathways: Cellular signaling routes (e.g., AMPK, mTOR, sirtuins) that respond to the presence or absence of nutrients and influence aging, longevity, and metabolic health.

Photodynamic Therapy: (Mentioned in some longevity contexts) Using light-activated compounds to target diseased cells—though not a primary focus here, any references align with emerging treatments to maintain skin and cellular health.

Placebo Effect: The phenomenon where a patient's positive expectation of treatment leads to perceived or real improvements in health, independent of the therapy's direct physiological effect.

Pluripotent Stem Cells (PSCs): Cells capable of giving rise to any cell type in the body. Induced pluripotent stem cells (iPSCs) are generated by reprogramming adult cells, offering a path to patient-specific regenerative therapies.

Proteostasis: The cellular quality control system that ensures proteins are properly folded and functional. Disruption in proteostasis leads to protein aggregation, contributing to diseases like Alzheimer's.

Rapamycin: A drug that inhibits mTOR, extending lifespan in animal models and currently under study for its potential to improve human healthspan.

Reactive Oxygen Species (ROS): Chemically reactive molecules containing oxygen, generated as byproducts of metabolism. While necessary in small amounts for signaling, excessive ROS can damage DNA, proteins, and lipids.

Resveratrol: A polyphenol found in red wine and grapes, studied for its potential to activate sirtuins and mimic some effects of caloric restriction, although human benefits remain debated.

Senescence: A state in which cells stop dividing and secrete harmful inflammatory signals (SASP), contributing to aging and age-related diseases.

Senolytics: Compounds designed to eliminate senescent cells, thereby improving tissue function and potentially extending healthspan.

Sirtuins: A family of proteins involved in DNA repair, metabolism, and aging regulation. Activating sirtuins may delay aging and extend lifespan.

Stem Cells: Undifferentiated cells capable of developing into various specialized cell types, crucial for tissue repair, regeneration, and potentially reversing age-related damage.

Telomerase: An enzyme that elongates telomeres, potentially reversing cell aging if safely activated, but with caution to avoid cancer risk.

Telomeres: Protective caps at the ends of chromosomes that shorten with each cell division, contributing to cellular aging once they become critically short.

Time-Restricted Feeding: A dietary approach limiting eating to a certain window of the day, aligning with circadian rhythms and offering potential metabolic and aging benefits.

Translational Research: Research that applies findings from basic science and animal studies to develop therapies, diagnostics, or technologies that improve human health.

Vaccines: Biological preparations stimulating an immune response to protect against infectious diseases; ongoing research explores their potential to prevent or mitigate age-related conditions.

Wearable Technology: Devices (e.g., smartwatches, fitness trackers) that continuously monitor health metrics (heart rate, sleep patterns, activity) to provide personalized feedback for longevity interventions.

Yamanaka Factors: A set of four transcription factors (OCT4, SOX2, KLF4, c-MYC) that can reprogram adult cells back to a pluripotent, embryonic-like state, a fundamental concept in epigenetic rejuvenation research.

REFERENCES

License Link References:

Creative Commons. (n.d.). *Attribution-ShareAlike 4.0 International (CC BY-SA 4.0) [License]*. Retrieved from https://creativecommons.org/licenses/by/4.0/ or https://creativecommons.org/licenses/by/2.0/

Web Sources/ Web Pages:

Worldometers. (n.d.). *World Population Statistics*. Retrieved from https://www.worldometers.info/world-population/

Gerontology Research Group. (n.d.). *World Supercentenarian Rankings List*. Retrieved from https://www.grg-supercentenarians.org/world-supercentenarian-rankings-list/

Lifespan Book. (n.d.). *Lifespan: Why We Age and Why We Don't Have To*. Retrieved from https://lifespanbook.com/

Life Force. (n.d.). *Tony Robbins: Get Started*. Retrieved from https://www.mylifeforce.com/

BrainyQuote. (n.d.). *Ray Kurzweil Quotes*. Retrieved from https://www.brainyquote.com/lists/authors/top-10-ray-kurzweil-quotes

PCMag News. (n.d.). *Understanding Epidemiology: A Comprehensive Guide*. Retrieved from https://www.pcmagnews.com/understanding-epidemiology-a-comprehensive-guide/

Horatio Alger Association of Distinguished Americans, Inc. (n.d.). *T. Denny Sanford*. Retrieved [Month Day, Year], from https://horatioalger.org/members/detail/t-denny-sanford/ Training, Coaching, Intervention Preeclampsia Georgia. https://www.thepreeclampsiaproject.com/training

Futuring Archives - RenAIssance Solutions. https://renaisol.com/category/futuring/

Vieten, C., Rubanovich, C., Khatib, L., Sprengel, M., & Tanega, C. (2024). Measures of empathy and compassion: A scoping review. PLoS One, 19(1), e0297099.

Musso, G., Gambino, R., & Cassader, M. (2022). The story of rapamycin (sirolimus): From the soil of Easter Island to a multitarget therapy. *Genes & Diseases, 9*(5), 1163–1178. https://doi.org/10.1016/j.gendis.2022.07.011Does Creatine Cause Hair Loss? | An Exploration of Hair Growth. https://evanalexandergrooming.com/blogs/the-den/does-creatine-cause-hair-loss

The Truth About HGH: Can It Really Improve Your Sex Drive? - HGH Vallarta. https://www.hghvallartaclinic.com/blog/the-truth-about-hgh-can-it-really-improve-your-sex-drive/

IGF-1 Bodybuilding: Unlocking the Potential -. https://tecamotest.com/igf-1-bodybuilding-unlocking-the-potential/

How Air.ai is scaling their integration marketplace with Cobalt. https://www.gocobalt.io/customer-stories/air-ai

How to design an office space for practicing mental health sessions. https://mentalhealthmarketing.com/how-to-design-an-office-space-for-practicing-mental-health-sessions/

Fight Aging. (2024, September). *Linking rapamycin, fasting, and spermidine in slowing aging*. Retrieved from https://www.fightaging.org/archives/2024/09/linking-rapamycin-fasting-and-spermidine-in-slowing-aging/

Loyal. (n.d.). HOMEPAGE. Retrieved December 30, 2024, from https://loyal.com/

The Benefits of Caffeine: More Than Just a Pick-Me-Up – No Nic Vapes – Go Nicotine Free in 2024. https://nonicvapes.com/the-benefits-of-caffeine-more-than-just-a-pick-me-up/

Unveiling the Health Benefits of Coffee: More than Just a Morning Boost - vasectomymedical.com. https://vasectomymedical.com/unveiling-the-health-benefits-of-coffee-more-than-just-a-morning-boost/

"The Evolving Cup: Unveiling Coffee and Decaf Trends in 2023 for a Healthier Lifestyle". https://www.interactivecrypto.com/magasine/the-evolving-cup-coffee-and-decaf-in-2023-healthier-than-we-thought

Dela Cruz, A. (2016). Evaluation of a Smoke and Tobacco-Free Initiative in a Student Wellness Center. https://doi.org/10.22371/07.2016.033

How Fiber Reduces Insulin Spikes – NUSTART. https://www.mynustart.com/blogs/insights/what-is-insulin-and-how-fiber-reduces-insulin-spikes-1

Is a Low Carb Diet Good for PCOS?. https://drbrighten.com/pcos-low-carb/

Understanding the Impact of Sugar on Weight Gain: Insights from the American Hospital Association - American Hospital Association's Physician Leadership Forum. https://www.ahaphysicianforum.org/health/understanding-the-impact-of-sugar-on-weight-gain/

Unlock Your Potential with Regenerative Medicine at Burick Center – Ask Us How! - Burick Center for Health and Wellness. https://burickcenter.com/unlock-your-potential-with-regenerative-medicine-at-burick-center-ask-us-how/

Musk, E. [@elonmusk]. (2024, April 15). [WE ARE ON THE EVENT HORIZON OF THE SINGULARITY] [Tweet]. X (formerly Twitter). https://x.com/elonmusk/status/1893810875875889507

Fox News. (2024, February 27). SILICON VALLEY ANTI-AGING INFLUENCER SAYS HE DOES NOT BELIEVE HE'LL DIE. Fox News. https://www.foxnews.com/media/silicon-valley-anti-aging-influencer-says-he-does-not-believe-hell-die.amp

American Parkinson Disease Association. (2025, May 14). New Parkinson's treatments in clinical trials. https://www.apdaparkinson.org/article/new-pd-treatments-clinical-trial-pipeline/

Arrazati, D. G. (2025, March 18). Longevity compound thwarts joint degeneration. NAD.com. https://www.nad.com/news/longevity-compound-thwarts-age-related-joint-degeneration

Bar-Ilan University. (2025, April 23). Evolutionary analysis uncovers protein in mammals. Phys.org. https://phys.org/news/2025-04-evolutionary-analysis-uncovers-protein-mammals.html

Cona, L. A. (2025). Stem cell therapy overview. DVC Stem. https://www.dvcstem.com/post/stem-cell-therapy

Cona, L. A. (2025). Stem cell treatment success rates. DVC Stem. https://www.dvcstem.com/post/stem-cell-success-rate

Dimension Market Research. (2025, January 30). Senolytic drugs market to reach USD 667.6 million by 2033. GlobeNewswire. https://www.globenewswire.com/news-release/2025/01/30/3018141/0/en/Senolytic-Drugs-Market-is-expected-to-reach-a-revenue-of-USD-667-6-Mn-by-2033-at-35-8-CAGR-Dimension-Market-Research.html

Flinders University. (2024, September 26). Circadian-informed lighting improves shift worker health. ScienceDaily. https://www.sciencedaily.com/releases/2024/09/240926131727.htm

Gameto. (2025, January 30). FDA IND clearance for Fertilo iPSC-based therapy. Business Wire. https://www.businesswire.com/news/home/20250130561032/en/Gameto-Announces-FDA-IND-Clearance-for-Fertilo-the-First-iPSC-Based-Therapy-to-Enter-U-S-Phase-3-Clinical-Trials

Hare, J., & Agafonova, N. (2025, March 12). Lomecel-B phase 2a trial results for Alzheimer's. BioSpace. https://www.biospace.com/press-releases/longeveron-announces-nature-medicine-publication-of-results-of-phase-2a-clinical-trial-evaluating-laromestrocel-lomecel-b-in-alzheimers-disease

Mohyeldeen, D. (2025, May 14). Time to see results from stem cell treatment. Beike Cell Therapy. https://beikecelltherapy.com/how-long-does-it-take-to-see-results-from-stem-cell-treatment/

National Institute on Aging. (2025, February 27). Senolytic therapy and bone health in women. https://www.nia.nih.gov/news/senolytic-therapy-shows-subtle-impact-age-related-bone-health-women

Neurona Therapeutics. (2025, February 20). NRTX-1001 reduces seizures in epilepsy patients. https://www.neuronatherapeutics.com/news/press-releases/022025/

NewLimit. (n.d.). Operating plan. Retrieved May 20, 2025, from https://www.newlimit.com/operating-plan

She, J.-X. (2025, March 12). NMN and its role in aging. Jinfiniti Precision Medicine. https://www.jinfiniti.com/what-is-nmn/

Thompson, B. (2024, July 18). Anti-aging potential of IL-11 inhibition. New Atlas. https://newatlas.com/medical/anti-aging-interleukin-11/

Vertex Pharmaceuticals. (2025, March 28). Program updates for type 1 diabetes portfolio. https://investors.vrtx.com/news-releases/news-release-details/vertex-announces-program-updates-type-1-diabetes-portfolio

Whitney, E. (2025, March 28). Thymosin beta-4 and TB-500 in aging. Innerbody. https://www.innerbody.com/thymosin-beta-4-and-tb-500

Journal Articles:

Gutiérrez, V., Monsalves, N., Gómez, G., Vidal, G., & Vidal, G. (2023). *Performance of a Full-Scale Vermifilter for Sewage Treatment in Removing Organic Matter, Nutrients, and Antibiotic-Resistant Bacteria. Sustainability, 15*(8), 6842.

Roth, S. M. (2008). *Perspective on the Future Use of Genomics in Exercise Prescription. Journal of Applied Physiology.* https://doi.org/10.1152/japplphysiol.01000.2007

Morales, M., Derbes, R., Ade, C., Ortego, J., Stark, J., Deininger, P., & Roy-Engel, A. (2016). *Heavy Metal Exposure Influences Double Strand Break DNA Repair Outcomes. PLoS One, 11*(3), e0151367.

Zhang, M., Schmitt-Ulms, G., Sato, C., Xi, Z., Zhang, Y., Zhou, Y., & Rogaeva, E. (2016). *Drug Repositioning for Alzheimer's Disease Based on Systematic 'omics' Data Mining. PLoS One, 11*(12), e0168812.

Ruby, J. G., Wright, K. M., Rand, K. A., Kermany, A., Noto, K., Curtis, D., ... & Jorde, L. B. (2018). Estimates of the heritability of human longevity are substantially inflated due to assortative mating. *Nature Communications,* 9(1), 1-10. https://doi.org/10.1038/s41467-018-02329-7

Arias, E., & Xu, J. (2022). United States Life Tables, 2019. *National Vital Statistics Reports, 70*(19), 1-33. National Center for Health Statistics. https://www.cdc.gov/nchs/data/nvsr/nvsr70/nvsr70-19-508.pdf

World Health Organization. (2020). World Health Statistics 2020. https://www.who.int/data/gho/publications/world-health-statistics

Barzilai, N., Atzmon, G., Schechter, C., Schaefer, E. J., Cupples, A. L., Lipton, R., ... & Shuldiner, A. R. (2003). Unique lipoprotein phenotype and genotype associated with exceptional longevity. *JAMA, 290*(15), 2030–2040. https://doi.org/10.1001/jama.290.15.2030

Atzmon, G., Schechter, C., Greiner, W., Davidson, D., Rennert, G., & Barzilai, N. (2004). Clinical phenotype of families with longevity. *Journal of the American Geriatrics Society, 52*(2), 274–277. https://doi.org/10.1111/j.1532-5415.2004.52070.x

What are the 4 types of marine resources? - Maritime Guide. https://maritime-union.org/what-are-the-4-types-of-marine-resources/

Look AHEAD Research Group. (2013). Cardiovascular effects of intensive lifestyle intervention in type 2 diabetes. *New England Journal of Medicine*, 369(2), 145–154.

U.S. Department of Health and Human Services. (2014). *The Health Consequences of Smoking—50 Years of Progress: A Report of the Surgeon General.*

Rehm, J. et al. (2010). The relation between different dimensions of alcohol consumption and burden of disease: an overview. *Addiction*, 105(5), 817–843.

Crous-Bou, M. et al. (2014). Mediterranean diet and telomere length in Nurses' Health Study: population based cohort study. *BMJ*, 349, g6674.

Ludlow, A. T. et al. (2013). Exercise training and metabolic syndrome in older individuals: a pilot study. *Physiological Reports*, 1(6), e00065.

Epel, E. S. et al. (2009). Can meditation slow rate of cellular aging? Cognitive stress, mindfulness, and telomeres. *Annals of the New York Academy of Sciences*, 1172, 34–53.

Messing, M., Torres, J., Holznecht, N., & Weimbs, T. (2024). Trigger Warning: How Modern Diet, Lifestyle, and Environment Pull the Trigger on Autosomal Dominant Polycystic Kidney Disease Progression. Nutrients, 16(19), 3281.

Collier, D. N., & Billings, C. J. (2007). Nonsteroidal Anti-inflammatory Drugs and Abdominal Pain. Pediatrics in Review. https://doi.org/10.1542/pir.28.2.75

Henriques, J. F. (2010). The Vibrations of Affect and their Propagation on a Night Out on Kingston's Dancehall Scene. Body & Society. https://doi.org/10.1177/1357034x09354768

Amor, C., Fernández-Maestre, I., Chowdhury, S., Ho, Y.-J., Nadella, S., Graham, C., ... Lowe, S. W. (2024). Senolytic CAR T cells reverse senescence-associated pathologies. Nature Aging, 4(1), 1–12. https://doi.org/10.1038/s43587-023-00560-5

Best, L., Dost, T., Esser, D., Flor, S., Mercado Gamarra, A., Haase, M., … Kaleta, C. (2025). Gut microbiota composition and its relation to human longevity. Nature Microbiology, 10(3), 1–15. https://doi.org/10.1038/s41564-025-01959-z

Brinton, E. A., Eckel, R. H., & Gaudet, D. (2025). Lipid-lowering therapies and cardiovascular health in aging. Atherosclerosis, 390, 117–125. https://doi.org/10.1016/j.atherosclerosis.2025.117615

Chen, L., Wu, B., Mo, L., Chen, H., Yin, X., Zhao, Y., … Tang, Y. (2025). Dietary interventions for metabolic aging. Nutrients, 17(6), 891–902. https://www.ncbi.nlm.nih.gov/pmc/articles/PMC11933296/

Chu, C., Wang, Y., Wang, Y., Fowler, C., Zisis, G., Masters, C. L., … Pan, Y. (2025). AI-driven prediction of Alzheimer's disease onset. JAMA Network Open, 8(1), e2435678. https://doi.org/10.1001/jamanetworkopen.2024.35678

de Lima Camillo, L. P., Asif, M. H., Horvath, S., Larschan, E., & Singh, R. (2025). Epigenetic clocks for aging assessment. Science Advances, 11(1), eadk9373. https://doi.org/10.1126/sciadv.adk9373

Dove, A., Wang, J., Huang, H., Dunk, M. M., Sakakibara, S., Guitart-Masip, M., … Papenberg, G. (2024). Diabetes, prediabetes, and brain aging. Diabetes Care, 47(10), 1794–1802. https://doi.org/10.2337/dc24-0862

Duarte Junior, M. A., Cabanas-Sánchez, V., Pintos-Carrillo, S., Ortolá, R., Rodríguez-Artalejo, F., Sotos-Prieto, M., & Martinez-Gómez, D. (2025). Mediterranean diet and longevity outcomes. The American Journal of Clinical Nutrition, 121(4), 567–575. https://doi.org/10.1016/j.ajcnut.2025.02.008

Eastwood, J., van Hemert, S., Stolaki, M., Williams, C., Walton, G., & Lamport, D. (2025). Gut microbiome and cognitive health in aging. The American Journal of Clinical Nutrition, 121(3), 345–356. https://doi.org/10.1016/j.ajcnut.2025.01.008

Fu, T. E., & Zhou, Z. (2025). Senolytic therapy with dasatinib and quercetin in aging models. Aging-US, 17(3), 123–134. https://www.ncbi.nlm.nih.gov/pmc/articles/PMC11921816/

Gao, C., Gong, N., Chen, F., Hu, S., Zhou, Q., & Gao, X. (2024). Marine-derived compounds for anti-aging applications. Marine Drugs, 23(1), 9. https://doi.org/10.3390/md23010009

Giovarelli, M., Zecchini, S., Casati, S. R., Lociuro, L., Gjana, O., Mollica, L., … De Palma, C. (2025). Stem cell therapies for muscle regeneration. Aging-US, 17(4), 456–467. https://www.ncbi.nlm.nih.gov/pmc/articles/PMC11977210/

Guo, Y., Yang, G., Liu, H., Chai, J., Chen, J., Shanklin, J., … Lu, M. (2025). Protein engineering for longevity pathways. Nature Communications, 16, Article 59549. https://doi.org/10.1038/s41467-025-59549-w

Hofer, S. J., Daskalaki, I., Bergmann, M., Friščić, J., Zimmermann, A., Mueller, M. I., … Madeo, F. (2024). Spermidine induces autophagy and extends lifespan. Nature Cell Biology, 26(8), 1345–1356. https://doi.org/10.1038/s41556-024-01468-x

Iqbal, T., & Nakagawa, T. (2024). NAD+ precursors and cellular aging. Biochemical and Biophysical Research Communications, 698, 149–156. https://doi.org/10.1016/j.bbrc.2024.149256

Kirkeby, A., Main, H., & Carpenter, M. (2025). Stem cell therapies for neurological disorders. Cell Stem Cell, 32(1), 45–56. https://doi.org/10.1016/j.stem.2024.00445

Landoni, J. C., Erkul, S., Laalo, T., Goffart, S., Kivelä, R., Skube, K., … Suomalainen, A. (2024). Mitochondrial function and aging. Nature Communications, 15, Article 52164. https://doi.org/10.1038/s41467-024-52164-1

Lelarge, V., Capelle, R., Oger, F., Mathieu, T., & Le Calvé, B. (2024). Senolytic interventions for bone health. npj Aging, 10, Article 138. https://doi.org/10.1038/s41514-024-00138-4

Liu, S., Faitg, J., Tissot, C., Konstantopoulos, D., Laws, R., Bourdier, G., … D'Amico, D. (2025). Urolithin A and muscle health in aging. iScience, 28(2), 107–114. https://doi.org/10.1016/j.isci.2025.108744

Loft, A., Emont, M. P., Weinstock, A., Divoux, A., Ghosh, A., Wagner, A., … Rosen, E. D. (2025). Brown adipose tissue and metabolic aging. Nature Metabolism, 7(5), 789–801. https://doi.org/10.1038/s42255-025-01296-9

Mahoney, S. A., Venkatasubramanian, R., Darrah, M. A., Ludwig, K. R., VanDongen, N. S., Greenberg, N. T., … Clayton, Z. S. (2024). Senolytic therapy and bone health in aging women. Aging Cell, 23(3), e14012. https://doi.org/10.1111/acel.14012

Maurer, S., Kirsch, V., Ruths, L., Brenner, R. E., & Riegger, J. (2025). Senolytic therapy restores chondrogenic phenotype. Osteoarthritis and Cartilage, 33(2), 234–245. https://doi.org/10.1016/j.joca.2024.10.007

Millar, C. L., Iloputaife, I., Baldyga, K., Norling, A. M., Boulougoura, A., Vichos, T., ... Lipsitz, L. A. (2025). Senolytic therapy for frailty reduction. eBioMedicine, 101, 104–112. https://doi.org/10.1016/j.ebiom.2025.104968

Morales, A. E., Dong, Y., Brown, T., Baid, K., Kontopoulos, D.-G., Ahmed, A.-W., ... Hiller, M. (2025). Genomic insights into bat longevity. Nature, 625(7993), 123–134. https://doi.org/10.1038/s41586-024-08471-0

Ocampo, A., Reddy, P., Martinez-Redondo, P., Platero-Luengo, A., Hatanaka, F., Hishida, T., ... Izpisua Belmonte, J. C. (2016). In vivo amelioration of age-associated hallmarks by partial reprogramming. Cell, 167(7), 1719–1733. https://doi.org/10.1016/j.cell.2016.11.052

Prokopidis, K., Moriarty, F., Bahat, G., McLean, J., Church, D. D., Patel, H. P. (2025). Nutritional interventions for healthy aging. Medicine, 104(7), e38210. https://doi.org/10.1097/MD.0000000000038210

Ruetz, T. J., Pogson, A. N., Kashiwagi, C. M., Gagnon, S. D., Morton, B., Sun, E. D., ... Brunet, A. (2024). CRISPR-based rejuvenation of neural stem cells. Nature, 627(8002), 345–356. https://doi.org/10.1038/s41586-024-07972-2

Santos-Gómez, A., Juliá-Palacios, N., Rejano-Bosch, A., Marí-Vico, R., Miguez-Cabello, F., Masana, M., ... Altafaj, X. (2025). Gene editing for neurological disorders. Molecular Genetics and Metabolism, 144(3), 123–130. https://doi.org/10.1016/j.ymgme.2024.101227

Senapati, P. K., Mahapatra, K. K., Singh, A., & Bhutia, S. K. (2025). mTOR-targeted therapies for aging. Biochimica et Biophysica Acta (BBA) - Molecular Basis of Disease, 1871(5), 167–174. https://doi.org/10.1016/j.bbadis.2025.167248

Song, P., Zhao, Q., & Zou, M.-H. (2020). Metformin and aging pathways. Aging-US, 12(10), 9876–9894. https://www.ncbi.nlm.nih.gov/pmc/articles/PMC7263313/

Sun, Q., Du, J., Tang, Y., Best, L. G., Haack, K., Cole, S. A., & Franceschini, N. (2025). Genetic variants in longevity pathways. JAMA Network Open, 8(3), e2431234. https://doi.org/10.1001/jamanetworkopen.2024.31234

Tessier, A.-J., Wang, F., Korat, A. A., Eliassen, A. H., Chavarro, J., Grodstein, F., ... Guasch-Ferré, M. (2025). Dietary patterns and healthy aging. Nature Medicine, 31(3), 456–467. https://doi.org/10.1038/s41591-025-03570-5

Tharmapalan, V., Du Marchie Sarvaas, M., Bleichert, M., Wessiepe, M., & Wagner, W. (2025). Epigenetic markers of aging. npj Aging, 11, Article 199. https://doi.org/10.1038/s41514-025-00199-z

Waziry, R., Ryan, C. P., Corcoran, D. L., Huffman, K. M., Kobor, M. S., Kothari, M., ... Belsky, D. W. (2023). Caloric restriction and aging biomarkers. Nature Aging, 3(2), 123–134. https://doi.org/10.1038/s43587-022-00357-y

Xia, J.-B., Liu, K., Lin, X.-L., Li, H.-J., Lin, J.-H., Li, L., ... Qi, X.-F. (2025). IL-11 inhibition extends lifespan in mice. Nature Communications, 16, Article 57962. https://doi.org/10.1038/s41467-025-57962-9

Xiong, X., Hou, J., Zheng, Y., Jiang, T., Zhao, X., Cai, J., ... Xie, C. (2024). Cellular senescence and neurodegenerative diseases. Cell Death & Disease, 15, Article 7062. https://doi.org/10.1038/s41419-024-07062-1

Xue, C., Yu, H., Pei, X., Yao, X., Ding, J., Wang, X., ... Guan, Y. (2025). Stem cell therapies for neurological repair. Neural Regeneration Research, 20(2), 345–356. https://doi.org/10.4103/1673-5374.39920784

Zumerle, S., Sarill, M., Saponaro, M., Colucci, M., Contu, L., Lazzarini, E., ... Alimonti, A. (2024). Senolytic therapies for cancer prevention. Nature Aging, 4(7), 987–998. https://doi.org/10.1038/s43587-024-00663-7

Villanueva, J. L., Adorno Vita, A., Zwickey, H., Fitzgerald, K., Hodges, R., Zimmerman, B., & Bradley, R. (2025). Dietary associations with reduced epigentic age. Aging-US. https://www.aging-us.com/article/206240/text

Preprints:

Blomquist, S. A., Kelly, G., Adães, S., Ardagh, A., Ramer, S., & Scuba, W. (2025). [Title of the preprint]. medRxiv. https://doi.org/10.1101/2025.03.19.25324259

Ercelen, D., Caggiano, C., Border, R., Sankararaman, S., Mangul, S., Zaitlen, N., & Thompson, M. (2024). [Title of the preprint]. bioRxiv. https://doi.org/10.1101/2024.11.30.625754

Macedo, O. C., da Silva, M. M., Magalhães, J. M., Sousa-Soares, C., Ala, M. I., Galhardo, M., ... Logarinho, E. (2025). Chemical enhancement of DNA repair in aging. bioRxiv. https://doi.org/10.1101/2025.02.21.639496

News Articles:

Blomquist, S. A., Kelly, G., Adães, S., Ardagh, A., Ramer, S., & Scuba, W. (2025). [Title of the preprint]. medRxiv. https://doi.org/10.1101/2025.03.19.25324259

Ercelen, D., Caggiano, C., Border, R., Sankararaman, S., Mangul, S., Zaitlen, N., & Thompson, M. (2024). [Title of the preprint]. bioRxiv. https://doi.org/10.1101/2024.11.30.625754

Macedo, O. C., da Silva, M. M., Magalhães, J. M., Sousa-Soares, C., Ala, M. I., Galhardo, M., ... Logarinho, E. (2025). Chemical enhancement of DNA repair in aging. bioRxiv. https://doi.org/10.1101/2025.02.21.639496

Reports/ Scientific Reports:

Horvath, S. (2013). *DNA Methylation Age of Human Tissues and Cell Types. Genome Biology.*

Campisi, J. (2011). *Cellular Senescence: Putting the Paradoxes in Perspective. Current Opinion in Genetics & Development.*

U.S. Department of Agriculture. (2024, December 10). Scientific report of the 2025 Dietary Guidelines Advisory Committee now available online [Press release]. https://www.usda.gov/about-usda/news/press-releases/2024/12/10/scientific-report-2025-dietary-guidelines-advisory-committee-now-available-online

World Health Organization. (2025, May 12). WHO results report 2024 shows health progress across regions overcoming critical challenges [News release]. https://www.who.int/news/item/12-05-2025-who-results-report-2024-shows-health-progress-across-regions-overcoming-critical-challenges

International Monetary Fund. (2025, April 16). How to build public support for energy subsidy and pension reforms [Blog post]. https://www.imf.org/en/Blogs/Articles/2025/04/16/how-to-build-public-support-for-energy-subsidy-and-pension-reforms

Informational Articles and Blog Posts:

LunaMD. (n.d.). *The Science Behind Aging: Genetic Factors.* Retrieved from https://lunamd.com/blogs/news/the-science-behind-aging-genetic-factors

Schneider, M. P., & Hilgers, K. F. (2017). *Specific Aldosterone Synthase Inhibition. Hypertension.* https://doi.org/10.1161/hypertensionaha.116.07939

Sinclair, D. A. (2015). *Sirtuins in Aging and Diseases: Past, Present, and Future. Cold Spring Harbor Perspectives in Medicine.*

Barzilai, N. (2018). *Exceptional Longevity: Insights from Centenarians. Annual Review of Medicine.*

Ocampo, A., et al. (2016). *In Vivo Amelioration of Age-Associated Hallmarks by Partial Reprogramming. Cell.*

Kirkland, J. L. (2017). *The Clinical Potential of Senolytic Drugs. Journal of the American Geriatrics Society.*

Gorbunova, V. (2012). *Cancer Resistance in the Blind Mole Rat is Mediated by Concerted Necrotic Cell Death Mechanism. Proceedings of the National Academy of Sciences.*

Scientific Advances and Blogs:

ScienceAlert. (n.d.). *Japanese Scientists Have Used Skin Cells to Restore a Patient's Vision for the First Time.* Retrieved from https://www.sciencealert.com/

Popular Mechanics. (n.d.). *Are Jellyfish Immortal? Facts About Jellyfish.* Retrieved from https://www.popularmechanics.com/

Mayo Clinic. (n.d.). *Spontaneous DNA Damage to the Nuclear Genome Promotes Senescence, Redox Imbalance, and Aging.* Retrieved from https://mayoclinic.pure.elsevier.com/

What is a Nutritious Serving of Meat? – Organic Prairie. https://www.organicprairie.com/blogs/news/what-is-a-nutritious-serving-of-meat

health - My Blog. https://trytada.com/review/health/

How Much Water Should You Drink Per Day? - Lake Oconee Health.
https://lakeoconeehealth.com/how-much-water-should-you-drink-per-day/
Educational and Research Websites:
CARTA. (n.d.). *Embryonic Stem Cell (ESC) | Center for Academic Research and Training in Anthropogeny.* Retrieved from https://anthropogeny.org/
AUVON Health. (n.d.). *New Frontiers in Stem Cell Therapy for Regenerative Medicine.* Retrieved from https://auvonhealth.com/
Mediluxe Medical Supplies. (n.d.). *Fagron TeloTest: A Comprehensive Guide to Telomere Testing for Aging Prevention.* Retrieved from https://en-us.mediluxegulf.com/
The Impact of Environmental Toxins on Weight: Insights from the American Hospital Association - American Hospital Association's Physician Leadership Forum.
https://www.ahaphysicianforum.org/health/impact-of-environmental-toxins-on-weight/
Brown fat is linked with lower risk of some chronic diseases..
https://podcast.foundmyfitness.com/news/s/epdxxo/brown_fat_is_linked_with_lower_risk_of_some_chronic_diseases
Armitage, H. (2020, April 6). 'Smart toilet' monitors for signs of disease. Stanford Medicine News Center. https://med.stanford.edu/news/all-news/2020/04/smart-toilet-monitors-for-signs-of-disease.html

Additional Resources:

BrainyQuote. (n.d.). Various Quotes by Authors and Thinkers. Retrieved from https://www.brainyquote.com/
Quotefancy. (n.d.). *Future quotes.* Retrieved from https://quotefancy.com/future-quotes
5 Signs You Might Need to Wind Down Your Wine Intake. https://thisnakedmind.com/wine-intake/
The Differences between Hemp and Marijuana - Silver Star Hemp. https://silverstarhemp.com/the-differences-between-hemp-and-marijuana/
Exploring the Impact of Cannabis on Liver Health in Adults. https://www.weedgets.com/en-se/blogs/newsletter/exploring-the-impact-of-cannabis-on-liver-health-in-adults
The Importance of B Vitamins in Hangover Recovery | Rebound.
https://reboundpartyrecovery.com/blogs/rebound-returns/why-you-need-b-vitamins-after-a-hangover
Aeschimann, W., Kammer, S., Staats, S., Schneider, P., Schneider, G., Rimbach, G., Cascella, M., & Stocker, A. (2021). Engineering of a functional γ-tocopherol transfer protein.
https://doi.org/10.1016/j.redox.2020.101773
Lebedev, M. A., & Nicolelis, M. A. L. (2017). Brain-machine interfaces: From basic science to neuroprostheses. *Progress in Brain Research, 218,* 1–64.
Martínez, V., & Sarter, M. (2008). Lateralized cognitive functions mediated by prefrontal cortex dopamine. *Neuropsychopharmacology, 33*(10), 2356–2371.
Polanía, R., Nitsche, M. A., & Ruff, C. C. (2018). Studying and modifying brain function with non-invasive brain stimulation. *Nature Neuroscience, 21*(2), 174–187.
Seo, D., Carmena, J. M., Rabaey, J. M., Alon, E., & Maharbiz, M. M. (2016). Neural dust: An ultrasonic, low power solution for chronic brain-machine interfaces. *arXiv preprint* arXiv:1307.2196.
Stocco, A., Prat, C. S., & Rao, R. P. (2014). Neuroscience: The brain in direct communication. *Scientific American, 311*(2), 26–29.
Gazzaniga, M. S. (2000). Cerebral specialization and interhemispheric communication: Does the corpus callosum enable the human condition? *Brain, 123*(Pt 7), 1293-1326.
Park, D. C., & Reuter-Lorenz, P. (2009). The adaptive brain: Aging and neurocognitive scaffolding. *Annual Review of Psychology, 60,* 173-196.
Small, G. W., Silverman, D. H. S., Siddarth, P., & Ercoli, L. M. (2006). Evaluating brain fitness activities: A clinical review. *Aging Health, 2*(3), 417-428.

In addition to source references, **Tad Sisler** drew from notes he took while reading books and publications and viewing conference presentations from the following leading scientists and individuals: **Dr. David Sinclair, Tony Robbins, Dr. Nir Barzilai, Dr. Vadim N. Gladyshev, Dr. Aubrey de Grey, Dr. Cynthia Kenyon, Dr. Brian Kennedy, Dr. Judith Campisi, Dr. Jan Vilg, Dr. Steve Horvath, Dr. Joao Pedro de Magalhaes, Dr. Felipe Sierra, Dr. Linda Partridge, Dr. Matt Kaeberlain, Dr. Luigi Fontant, Dr. Eric**

Verdin, Dr. Thomas Rando, Dr. Valter Longo, Dr. James Kirkland, Dr. Peter de Keizer, Dr. Manuel Serrano, Dr. Peter Attia, Ilchi Lee, Steven Gundry, M.D., Peter H. Diamandis, Steven Kotler, and Ray Kurzweil.

Tad also researched the most current advancements in longevity and age reversal he could find from these and other notable laboratories: *Sinclair Lab at Harvard Medical School, Altos Labs, Calico Life Sciences, Buck Institute for Research on Aging, Salk Institute for Biological Studies, Potocsnak Longevity Institute at Northwestern University, Barshop Institute for Longevity and Aging Studies, Sanford Center on Longevity, Juvenescence,* and *Unity Biotechnology.*

Tad also included ideas and the most current breakthroughs from his notes from publications in peer-reviewed journals, including *The New England Journal of Medicine, Ageing Research Reviews, Aging Cell, The Journals of Gerontology, Series A: Biological Sciences and Medical Sciences, Biogerontology,* and *Rejuvenation Research.* In addition, **Tad** included beneficial information for longevity and physical and mental health from the teachings of seers and legendary thinkers, including **Edgar Cayce, Hippocrates, Paracelsus, Leonardo da Vinci, Nostradamus, St. Hildegard of Bingen, Sir Francis Bacon, Dr. Serge Voronoff, Carl Jung, Nikola Tesla, Pythagoras, Ayurvedic Charaka, Sushruta Sages, and Elon Musk.**

For the most up-to-date information, readers are encouraged to consult recent publications and ongoing clinical trials in the field of aging research.

ABOUT THE AUTHOR

Tad Sisler is an American Composer, Author and Producer of feature films and music. More than a thousand of his original works are available through *iTunes, Amazon* and virtually every other major marketplace. Through the years, **Tad** created and released independent feature films and documentaries, television shows, developed a music store and vast collection of music for film and television usages, in addition to published screenplays and books.

Tad is a voting member of *The Academy of Recording Arts & Sciences.* **Tad** invented a wireless karaoke all-in-one microphone that became a best-seller on *Amazon.* A child prodigy, Tad was playing advanced piano pieces at the age of 8, and rating superior in Classical piano competitions at 12. Tad won his first scholarship for singing at 12, attending the Idyllwild School of Music and the Arts, then affiliated with the University of Southern California.

FEATURE FILMS
Tad produced, edited, and released "**The Ghosts of Brewer Town**", a mystery feature film, currently available on *YouTube.*

TELEVISION PROJECTS
Tad launched the **Journey To An Extraordinary Life-Legends Among Us** documentary series, which chronicles the lives and careers of legendary artists, actors, sports figures and heroes of medicine, in a feature-film format.

BOOKS
Books, Audio Books and Podcasts released by **Tad** include "**Reflections in the Key of Life-The Steve Madaio Story**", chronicling the life and times of America's most prolific trumpeter. This book garnered a **Readers' Favorite Book Award** for Tad.

"**Mafia Baby**" is a shocking true story of a woman raped by a Mafioso, who then raised his child alone. Tad's autobiography, "**It's a Long Climb to The Middle**" *is* available currently on *Amazon* and *Barnes & Noble.* Screenplays in development by Tad Sisler include "**The Incredible Spark of Franklin Benjamin**", and "**Please Don't Forget**". Tad's latest **Music Mastery** collection of books is designed to educate and inspire musicians to become masters. His **Health and Longevity Mastery** series of books is crafted to educate on longevity, age reversal, and general wellness.

MUSIC

Tad's production music catalog tripled in size with the addition of thousands of excellent production music tracks, as well as hundreds of sound-alike tracks for the DJ/Karaoke industry, now distributed on **iTunes, Amazon Marketplace, CD Baby, Spotify, Rdio, Xbox Music** and dozens of other outlets Worldwide.

Tad produced and released "The Barcelona Sessions" to 1000 radio stations Worldwide, with never-before-heard original performances by Miles Davis' bassist, Bill Evan's drummer, Frank Sinatra's saxophonist, Maynard Ferguson's guitarist, and Andrae Crouch' flutist/saxophonist, produced by Tad Sisler in his recording studio.

Tad Sisler composed the full score to **"The Encore Of Tony Duran"**, an indie feature film starring **Elliott Gould, William Katt, Nicki Ziering and Cody Kasch**, along with his co- composer Andrew Fraga, Jr. After having the distinction of being the first film to sell-out at the prestigious *Palm Springs International Film Festival*, the film won the **Jury Award** for **Best Feature Film** at the *Las Vegas Film Festival* and the *Santa Fe Film Festival*, as well as the **Indie Spirit Award** at the *Fort Lauderdale Film Festival* and the **Audience Favorite Award** at *Tallgrass Film Festival*, in conjunction with a **Lifetime Achievement Award** for **Elliott Gould.** The film is available on *Amazon Prime*.

Tad completed the music and audio editing for the TV Series **"American M.C.".** The first 7 episodes are complete and in the process of distribution through **iTunes**. Tad scored the Main Title theme to **American M.C.** as well as underscore and providing Music Supervision and source music.

PRODUCTION

Tad Sisler has been a valuable member of the team of specialists and project developers for **Yamaha Corporation of America**, delivering hundreds of intricate projects to exact **Yamaha** specifications over a 10-year period.

Tad received accolades in 2011 after being given the honor and challenge of doing the "official" remake of the iconic **"Andy Griffith Theme"** for the estate of the composer **Earle Hagen** as a perfect sound-alike, along with his composing associate Andrew Fraga, Jr.

Following a stint composing for a series entitled **"Famous Families"** on **Foxstar** and working as assistant to composer Jeff Edwards on the television series **"Silk Stalkings"** and **"Renegade"** in the late 1990's, Tad Sisler and founded & developed a production music catalog, containing thousands of high-quality music tracks available for sync licenses in film, television and advertising in more than 150 genres.

In addition to handling Music Supervision on **"The Encore Of Tony Duran"**, and on **"American M.C.", "The Ghosts of Brewer Town", "Tis' The Season"**, the **"Journey To an Extraordinary Life"** series, **Tad** placed his original music on **NBC, ABC/Disney, Warner Brothers Television, TNT**, US National Infomercial campaigns through **Guthy/Renker** and **Script To Screen**, as well as custom composing for the TV and Advertising industry.

Tad released contains hundreds of top-quality soundalike tracks produced by **Tad** and his associates, for DJ and Karaoke usages, currently on *ITunes, Amazon Marketplace, Spotify, Rdio, Xbox Music,* and many other outlets.

LIVE PRODUCTION

In the 1980's and 1990's, **Tad** and his team produced a series of live headliner events at multiple venues from the ground up, including sold-out performances by **Kenny Rogers, Earth, Wind & Fire, Los Lobos, Glen Campbell, The Righteous Brothers, Lou Rawls, Tito Puente,** the **Power Jam** featuring **Timmy T, Tara Kemp, Candyman, Soul To Soul** and more.

HISTORY

As a very young man, Tad Sisler worked as a performer for **Frank Sinatra**, studied music in choreography under world-famous Broadway Dancer/Choreographer **Jacque D'Amboise**, received superior ratings in classical piano performance in tough **Joanna Hodges** international competitions, and received private acting lessons from **Richard Burton**, a friend of his family.

Tad attended the prestigious **Idyllwild School of Music and the Arts** on vocal music scholarships during the period when it was affiliated with the **University of Southern California**. In High School, Tad was one of 100 statewide vocalists elected to the prestigious **All-State Choir** in Missouri.

During his storied career, Tad has also had the honor of performing with and working among such greats as **Gladys Knight, Rita Coolidge, B.B. King, Marilyn McCoo, Johnny Mathis, Kenny Rogers, Tito Puente, Sonny and Mary Bono, Gene Barry, Terry Cole-Whittaker, Shecky Greene, Peter Marshall, Mary Hart, Blackwell, Herb Jeffries, Trini Lopez, Glen Campbell, Jennifer Hudson** and other legends.

Tad Sisler's extensive experience, state of the art facility and history of delivering quality feature films and music on time and on budget, as well as the ability to draw from an extensive catalog of production music, allows his experienced team to offer complete services in custom film and television production as well as in music composition and production efficiently.

Tad is proud and humbled to be a voting member of the **Academy of Recording Arts & Sciences**, which allows him to have a voice to vote for great artists worthy of winning a **Grammy Award**. Many of Tad's works have been placed into Grammy consideration.

In 2023, Tad won a prestigious **Telly Award** for creative excellence in his *Journey to an Extraordinary Life* film series.

Modern Renaissance Publishing is at the forefront of a new intellectual awakening, dedicated to fostering a renaissance of ideas that resonate in today's world. Our mission is to bring cutting-edge concepts and timeless wisdom to the public through a diverse array of publishing formats, including books, eBooks, and audiobooks.

We are proud to launch our **Music Mastery** series, offering comprehensive guides and insights for musicians of all levels. Our **Health and Longevity Mastery** series highlights the latest discoveries and insights into extending the human healthspan and lifespan. In addition to our literary endeavors, we also publish original music, enriching the cultural landscape with creative expressions. Whether you're seeking to expand your knowledge, enhance your skills, or simply be inspired,

Modern Renaissance Publishing provides the resources and content to empower your journey. Join us as we bridge the rich heritage of the past with the innovative spirit of the present to shape a brighter, more enlightened future.

©2025 by Tad Sisler
Publisher: MODERN RENAISSANCE PUBLISHING
IN USA +1 (818) 845-6700
modernrenaissancepublishing.com
Email: modernrenaissancepublishing@gmail.com
ISBN# 978-1-966258-17-9

MODERN RENAISSANCE
PUBLISHING

www.ingramcontent.com/pod-product-compliance
Lightning Source LLC
Chambersburg PA
CBHW060226030426
42335CB00014B/1352